The Mother Lode Gold System of California

by Adolph Knopf

with an introduction by Kerby Jackson

Introduction

It has been ninety years since the USGS released their important publication "The Mother Lode System of California". First released in 1929, this important volume has now been out of print for nearly a century and has been unavailable to the mining community since those days, with the exception of expensive original collector's copies and poorly produced digital editions.

It has often been said that "*gold is where you find it*", but even beginning prospectors understand that their chances for finding something of value in the earth or in the streams of the Golden West are dramatically increased by going back to those places where gold and other minerals were once mined by our forerunners. Despite this, much of the contemporary information on local mining history that is currently available is mostly a result of mere local folklore and persistent rumors of major strikes, the details and facts of which, have long been distorted. Long gone are the old timers and with them, the days of first hand knowledge of the mines of the area and how they operated. Also long gone are most of their notes, their assay reports, their mine maps and personal scrapbooks, along with most of the surveys and reports that were performed for them by private and government geologists. Even published books such as this one are often retired to the local landfill or backyard burn pile by the descendents of those old timers and disappear at an alarming rate. Despite the fact that we live in the so-called "Information Age" where information is supposedly only the push of a button on a keyboard away, true insight into mining properties remains illusive and hard to come by, even to those of us who seek out this sort of information as if our lives depend upon it. Without this type of information readily available to the average independent miner, there is little hope that our metal mining industry will ever recover.

This important volume and others like it, are being presented in their entirety again, in the hope that the average prospector will no longer stumble through the overgrown hills and the tailing strewn creeks without being well informed enough to have a chance to succeed at his ventures.

Kerby Jackson
Josephine County, Oregon
October 2015

CONTENTS

	Page
Outline of the report	VII
Introduction	1
Geography	1
Situation of the district	1
Physical features	1
Climate and vegetation	1
Field work and acknowledgments	3
Previous geologic studies of the district	3
Present investigation	3
Historical sketch	4
Output of gold	7
General geology of the Mother Lode belt	8
Geology of the lower western slope of the Sierra Nevada	8
Detailed features of the geology of the Mother Lode belt	10
Calaveras formation	10
Amphibolite schist associated with the Calaveras formation	11
General features	11
Age	12
Structure	12
Mariposa slate	12
General character	12
Petrographic features	13
Conglomerate	13
Age	14
Greenstones associated with Mariposa slate	14
General features	14
Augite basalts (breccia, tuffs, and lavas)	15
Keratophyric rocks	16
Contemporaneous intrusions	17
Conclusion	18
Intrusive igneous rocks	18
General relations	18
Metadiorite	18
Igneous rocks of post-Mariposa age	19
Serpentine	19
Hornblendite and gabbro	20
Granodiorite	21
Albitite and albitite porphyry	21
General features	21
Petrography	21
Geothermal gradient	22
Gold ore bodies	23
General description	23
Water conditions underground	24
Quartz veins	24
General features	24
Vein filling	25
Gouge	26
Ore shoots	26
General features	26
Distribution of the gold within the shoots	27
Origin of the ore shoots	29
Previous ideas	29
Influence of wall rocks	31
Proposed hypothesis	31
Ore bodies of mineralized country rock	33
Mineralized greenstone ("gray ore")	33
Nature and occurrence	33
Origin	33
Mineralized schists	34

CONTENTS

	Page
Mineralogy	34
Alteration of wall rocks by the ore-forming solutions	41
Alteration of the Mariposa slate	41
Alteration of graywacke, conglomerate, and quartzite	43
Alteration of the greenstones	43
Alteration of the amphibolite and allied schists	43
Alteration of serpentine	44
Alteration of chlorite schist (sheared pyroxenite)	44
Summary of wall-rock alteration	44
Origin of the gold deposits	45
The mines	48
Plan of treatment	48
Georgia Slide mine	48
Guildford mine	49
Plymouth mine	49
History	49
Development	50
Output	50
Geology	51
Fremont mine	52
Treasure mine	54
Bunker Hill mine	55
Original Amador mine	56
Keystone mine	58
Eureka mine	59
Central Eureka mine	59
South Eureka mine	62
Kennedy mine	63
Argonaut mine	66
Moore mine	68
Mammoth mine	69
Amador Queen No. 2 mine	70
Ford mine	70
Union mine	71
Gold Cliff mine	71
Royal mine	72
Mines on Carson Hill	72
History	72
General geology	73
Ore bodies	73
Norwegian mine	77
Chilena mine	78
Dutch Sweeney and App mines	78
Eagle Shawmut mine	79
Clio mine	82
Mariposa estate	83
History and general features	83
Queen Specimen mine	84
Princeton mine	84
Index	87

ILLUSTRATIONS

	Page
PLATE 1. *A*, Mother Lode belt, as seen from the Central Eureka mine; *B*, Mother Lode belt, as seen from summit of Carson Hill	1
2. Geologic map of the Mother Lode belt, California	In pocket.
3. Location of Mother Lode belt and other mining districts with reference to the intrusive rocks of the Sierra Nevada	8
4. *A*, Ribbon quartz, Plymouth mine; *B*, Inclusions of greenstone and black slate in quartz, Original Amador mine	24
5. *A*, Composite quartz ore from the 4,200-foot level, Argonaut mine; *B*, Prominent quartz outcrop at Eagle Shawmut mine	25
6. Distribution of gold and structural features of typical quartz vein inclosed in slate, Plymouth mine	28
7. *A*, Ankerite metasome in slate, ordinary light; *B*, Same as *A*, but in polarized light; *C*, Metasomatic ankerite in slate, Sherman mine	40

CONTENTS

Page

PLATE 8. *A*, Ankerite metasomes in slate, surrounded by fringes of clear ankerite; *B*, Metasomatic pyrite in slate surrounded by feather quartz_____ 41
 9. Geologic map of the area surrounding the Plymouth mine _____ 56
 10. Ore shoot above the 4,100-foot level, Central Eureka mine_____ 56
 11. Stereogram of ore shoots of Morgan and Melones mines_____ 80
 12. Geologic map of the adit level, Eagle Shawmut mine_____ 80
FIGURE 1. Index map showing the location of the Mother Lode belt, California_____ 2
 2. Output of Mother Lode counties, 1904–1924_____ 9
 3. Geologic map of Mother Lode belt from Plymouth to Jackson_____ 17
 4. Structural features of a quartz vein in slate_____ 26
 5. Thickness and assay value of ore shoot on the 2,150-foot level, Plymouth mine_____ 28
 6. Concentration of gold in the ore shoot on the 2,150-foot level, Plymouth mine_____ 29
 7. Section through the Pacific shaft, Plymouth mine_____ 52
 8. Ore shoots of Plymouth mine_____ 53
 9. Section through Gover shaft, Plymouth mine_____ 54
 10. Section through Bunker Hill mine 800 feet north of the shaft_____ 56
 11. Section through the main shaft of the Original Amador mine_____ 57
 12. Section through the Patton shaft of the Keystone mine_____ 58
 13. Section through the shaft of the Central Eureka mine_____ 60
 14. Detailed geologic cross section of the lower levels of the Central Eureka mine._____ 61
 15. Generalized section through the Kennedy mine along the line of the 1,950-foot crosscut_____ 64
 16. Section through the Kennedy vein 280 feet north of the vertical shaft, showing the branching of the vein_____ 64
 17. Stope plan of the Kennedy vein on the 3,900-foot level_____ 65
 18. Section through the Argonaut shaft_____ 67
 19. Geologic map of the 3,300-foot level of the Argonaut mine_____ 68
 20. Section through the Moore mine on a line parallel to and 60 feet south of the 640-foot west crosscut_____ 69
 21. Pocket veins at the Mammoth mine_____ 70
 22. Plan of part of the 1,600-foot level, Melones mine, showing the divergence in strike between the Bull vein and the country rocks_____ 75
 23. Partial section through the Melones mine along the line of the lower winze on the 1,100-flat vein_____ 76
 24. Vertical section through the 610-foot winze, Eagle Shawmut mine_____ 80
 25. Longitudinal section through the Eagle Shawmut mine, showing the distribution of the ore by kinds_____ 81
 26. Section through the Clio shaft_____ 83

OUTLINE OF THE REPORT

The Mother Lode belt of California is a long, narrow strip on the western foothills of the Sierra Nevada, in which is inclosed a more or less continuous series of gold deposits. It is a mile wide and 120 miles long. Because of the great wall-like masses of quartz that crop out at intervals along the belt, there grew up in the early fifties the belief that a quartz vein extends continuously from one end of the belt to the other—the Mother Lode. As the gold deposits are not continuous and do not consist of quartz veins only, the term Mother Lode system is more appropriate. The variety of gold deposits that occur in the Mother Lode system is very considerable, and therein lies one of the possibilities of future expansion of output, for attention has perhaps been focused too sharply on the quartz-filled fissures as the main mode of occurrence of the gold.

The total output to the end of 1924, is estimated to be $240,000,000. Half of this output has come from a 10-mile segment of the belt, extending from Plymouth to Jackson, in Amador County.

Mining began in 1849, on the Mariposa grant at the south end of the belt. The vertical depth now attained is 4,500 feet; the ore at that depth is on the average as good as that mined on the upper levels, and it is thought that mining can be profitably extended to a vertical depth of 6,000 feet. Mining on the Mother Lode belt is favored by nearness to supply centers, by good transportation facilities, labor supply, and climatic conditions, and by the slow increase of rock temperature in depth. These advantages are partly offset in the mining of the quartz veins by the moderate gold content of the ore, the moderate size of the ore bodies, and the high cost of timbering, due to heavy ground.

The rocks traversed by the Mother Lode system consist mainly of steeply dipping slates, schists, and greenstones, and include some intrusive masses of serpentine. They form a series of belts that trend northwest, parallel to the course of the range.

The oldest rocks belong to the Calaveras formation, of Carboniferous age and of sedimentary origin. They consist of black phyllites, not far removed from black slates in appearance, with subordinate quartzite and limestone. Associated with the Calaveras formation are green schists, which are of particular interest, for many of the noted mines of the Mother Lode system occur in them. They generally contain chlorite or amphibole, or both, to which they owe their prevailing gray-green color. In places they are interbedded with a phyllite of the Calaveras formation. They have been derived by metamorphism from augitic tuffs and lavas erupted in Calaveras time.

The Calaveras rocks were metamorphosed, probably at the end of the Paleozoic era, and after the metamorphism igneous magmas invaded them, yielding diorite, quartz diorite, muscovite granite, aplite, and hornblende lamprophyre.

The youngest stratified rocks of the belt make up the Mariposa slate. Black slate with less graywacke and minor conglomerate and tillite (?) are the normal members of the formation. Greenstone, chiefly in the form of tuff and breccia, is associated with the Mariposa formation; it occurs in thick belts with little or no intercalated slate, and it is also intimately interbedded with Mariposa slate and graywacke. Moreover, tuffaceous material is mixed with the normal sedimentary material, forming mixed rocks that range from nearly pure argillaceous slate to pure greenstone.

The associated greenstones were termed augite porphyrite and diabase by Lindgren and Turner, and later they were called meta-andesites by Ransome. However, the nature of few if any of the feldspars in these rocks has been recorded. Along the Mother Lode belt the feldspars prove generally to be albite. At present it is an open problem how much of the abundant albite in the thick belts of volcanic greenstone is the result of albitization of calcic plagioclase and how much is pyrogenic; hence the rocks will here be referred to by the noncommittal terms greenstone and augite melaphyre. Basalts, albitized augite melaphyres, and keratophyres are recognized.

In late Jurassic or early Cretaceous time the region of the Sierra Nevada felt the onset of the Cordilleran revolution. The rocks were isoclinally folded and then were invaded by a succession of plutonic rocks, beginning with peridotite, which was soon altered to serpentine, and ending with granodiorite. The remarkable albitite dikes near Tuolumne River also were intruded probably at this time. At the end of this epoch of igneous activity the gold deposits were formed. Since the Cordilleran revolution the gold belt has remained above sea level; and during that long span of time erosion has stripped off several thousand feet of rock.

The northern part of the Mother Lode belt as far south as the Gwin mine in Calaveras County is inclosed in alternating Mariposa slate and associated greenstone. From the Gwin mine southward as far as Tuolumne River the lode is chiefly in green schist; it there enters serpentine, in which it continues to Coulterville, where it enters the Mariposa slate, and it continues in this formation until it is terminated by the granodiorite south of Mariposa. In this characterization details have been omitted. The differences in geologic environment just sketched determine certain differences in the gold deposits that the rocks inclose.

The gold deposits are of two principal kinds—quartz veins and bodies of mineralized country rock. Although the quartz veins, especially those that are inclosed in slate, are similar from one end of the belt to the other, the ore bodies of mineralized country rock show far more individuality and more geologic complexity than was expected. The statistics of output show that there has been a marked localization of gold production in certain portions of the belt, notably the 10-mile segment in Amador County, Angels Camp, and Carson Hill in Calaveras County, and a segment south of Jamestown, Tuolumne County.

The ores are of low or moderate grade, averaging $7 a ton. The very low-grade ores formerly mined, ranging from $2 to $3 a ton, are no longer worked. The gold bullion ranges in fineness from 790 to 840 parts per thousand, but the gold as it occurs in the ores ranges from 839 to 899 parts, its fineness as determined on the few samples analyzed depending apparently on the nature of the associated sulphide or telluride. Sulphides make between 1 and 2 per cent of the quartz ore and double or triple that amount in the ore bodies of mineralized country rock. They consist almost wholly of pyrite, with minor arsenopyrite, sphalerite, galena, chalcopyrite, and tetrahedrite. The telluride petzite appears to be restricted to the portion of the Mother Lode in Calaveras and Tuolumne Counties. Arsenopyrite is fairly common in Amador County. Galena and petzite indicate good ore, but the others are indifferent indicators.

The quartz veins generally occur as systems of parallel or acutely intersecting veins; as many as four may be worked at a single mine. Not many of them can be definitely traced for more than a few thousand feet. They cut the inclosing rocks at an acute angle in both strike and dip. They fill fissures that were formed by reverse faulting, in some of which the displacement amounts to 375 feet. Structurally they do not conform to the idea that veins follow lines of least resistance; they make an acute angle with the well-developed cleavage of the slate, and they leave slate to pass into massive greenstone, and vice versa. Where this happens the vein is markedly deflected or refracted, and this refraction is so conspicuous a structural feature of the veins of Amador County that a rough value for the index of refraction was computed—1.4.

The veins swell and pinch abruptly; in the lenticular expansions the filling is quartz; at the edges of lenses, either along the strike or on the dip, the massive quartz filling becomes more and more admixed with slate or pinches down to a gouge-filled fissure. Banding or ribboning is common in veins that are inclosed in slate or schist and is invariably parallel to the walls of the veins; locally the divergence between ribboning and bedding is as much as 45°.

The ore occurs in shoots generally short but persistent in depth. The shoots as a rule have a steep pitch or "rake" which may be either north or south. The ore shoots are considerably wider than adjacent portions of the vein, as well as containing more gold to the ton. Furthermore, the shoots consist more solidly of quartz, and where the massive quartz filling of the ore shoot frays out into a stringer lode it becomes poor in gold. Unfortunately, however, large bodies of quartz do not necessarily make ore even if they occur along fissures that elsewhere contain ore shoots. Most of the known ore shoots cropped out at the surface but some have been discovered whose tops were as deep as 3,300 feet. Intersection and branching of veins favored the development of ore shoots. Not all junctions, however, have determined ore shoots.

The gold is believed to have come from a deep-seated magmatic source, the quartz came largely from the adjacent wall rock, and the water by which these substances were carried in solution came partly from the magma and largely from the meteoric circulation. As the thermal water rising in a fissure issues at the earth's surface only at certain orifices, such localized efflux would determine more rapidly moving threads of current in the fissure; all other water in the fissure would be stagnant or nearly stagnant. Only those portions of the fissure in which the ore-depositing solution was actively flowing would become the loci of gold deposition.

It is unsafe to generalize for the whole Mother Lode belt as to the influence of the nature of the wall rocks on the tenor of the ore. Valuable ore bodies have formed in rocks of many kinds. Probably the only valid generalization is the one made by Ransome that paying veins may occur in any rock with the possible exception of serpentine. However, the slate appears to be more favorable than the greenstone, and veins wholly inclosed in greenstone are likely to be of low grade. Although the favorable character of the slate has been ascribed to its carbonaceous content, the real explanation may be that the slate delivered quartz less rapidly to the veins than the greenstone or other rocks; the ore, so to speak, was less diluted with quartz.

The ore bodies of mineralized country rock comprise the so-called gray ore and the mineralized schist. They occur either adjacent to quartz veins or in broad zones of fissuring. The gray ore is the result of hydrothermal attack on the augitic greenstone; it consists of ankerite, sericite, albite, quartz, pyrite, and generally some arsenopyrite and contains enough gold to be profitably workable. Many of the shoots were large and averaged as high as $8 a ton. These shoots generally adjoin thin quartz veins or are in the wedge between two intersecting

veins. The tenor of the gray ore can be determined only by assays. There may well be large deposits of gray ore undiscovered, but guides to their discovery are unknown.

The mineralized schist ore bodies consist of pyritic sericite-ankerite schist and are the results of the alteration of amphibolite and chlorite schists. They are generally ramified with quartz-ankerite veinlets. They are of low grade as a rule, carrying from $2 to $3 a ton, but one of the highest-grade ore bodies ever mined on the Mother Lode belt, the hanging-wall ore body of the Morgan-Melones mine, on Carson Hill, is a pyritic ankeritized schist. Many of these ore bodies form either the footwall or the hanging wall of a large thick barren quartz vein. At Carson Hill the schists in the footwall of the thick barren quartz vein made a very low-grade ore body, whereas the schists in the hanging wall made the high-grade ore body already mentioned.

The wall rocks of the Mother Lode veins have been profoundly altered by the ore-forming processes. Large volumes of rock have thereby been transformed. Carbonatization (ankeritization) was the chief effect, regardless of whether the rocks were slate, graywacke, quartzite, conglomerate, greenstone, amphibolite schist, chlorite schist, talc schist, or serpentine. Sericite, albite, pyrite, and arsenopyrite were also commonly formed by chemical attack on the wall rocks. Gold also migrated into the wall rocks, as already described.

Serpentine and the augitic greenstones were the most susceptible to ankeritization. The ankeritized serpentine, which is particularly distinctive of the southern part of the lode, makes up belts hundreds of feet thick, generally tinted a delicate green by the presence of the chromiferous potassium mica mariposite. The black slates because of their fine grain and carbonaceous pigment seem to be unaffected, but microscopic and chemical analyses prove that they too have been greatly altered, chiefly by ankeritization. Moreover, the microscope shows that the ankerite augen that have formed in the slates by replacement have been partly rotated by the compressive forces by which the vein fissures were produced.

Great quantities of carbon dioxide were added to the wall rocks, and correspondingly great quantities of silica were eliminated from them. This silica was more than enough to supply the quartz in the Mother Lode veins, and this fact leads to the surmise that the somewhat unfavorable effect of serpentine and greenstone on the veins in them may have been due to the rapidity with which they supplied quartz to the growing veins.

The Mother Lode fissure system, according to the hypothesis presented in this report, appears to occupy a series of auxiliary fractures in a zone parallel to a great reverse fault. Facts in support of this hypothesis are so far known only from that part of the belt extending from Plymouth to Jackson.

The veins can best be explained as having grown by successive enlargement along fissures that were from time to time reopened by renewal of movement along them. To some extent the force of crystallization of the growing quartz may have aided in producing the ribbon structure of the veins. The quartz of the veins was supplied from the wall rocks; the gold, sulphur, arsenic, carbon dioxide, and certain other constituents were probably supplied by exhalations that issued at a high temperature from a deep-seated consolidating granitic magma, as was also a part of the thermal energy of the ore-forming solutions. After these exhalations had condensed to water that carried the other constituents dissolved in it, the motive power for causing this "magmatic water" to rise was doubtless the gravity potential of a meteoric circulation whose paths were determined by the fissure systems. These processes took place near the end of the epoch of plutonic intrusion that marked the final stage of the Cordilleran revolution in late Jurassic or early Cretaceous time.

THE MOTHER LODE SYSTEM OF CALIFORNIA

By Adolph Knopf

INTRODUCTION

GEOGRAPHY

SITUATION OF THE DISTRICT

The Mother Lode belt of California is a strip a mile or so wide extending for 120 miles along the lower western flank of the Sierra Nevada. It begins near Georgetown, in Eldorado County, and extends to Mormon Bar, 2 miles south of Mariposa, in Mariposa County. (See fig. 1.) The five counties that it traverses—Eldorado, Amador, Calaveras, Tuolumne, and Mariposa—are often known as the Mother Lode counties.

The main towns on the belt from north to south are Placerville, Plymouth, Sutter Creek, Jackson, San Andreas, Angels Camp, Jamestown, and Mariposa. Placerville, with a population of 1,650, is the largest, but its prosperity is no longer based on placer mining; it now rests mainly on horticulture. Plymouth and Sutter Creek are dependent chiefly on quartz mining, but the other towns are largely independent of mining. Jackson is the seat of Amador County, though its fortunes are closely linked to those of the two great gold mines near it—the Argonaut and the Kennedy. San Andreas is the seat of Calaveras County. Angels Camp, in the same county, is in a flourishing condition, though all its mines are idle; it is hard to realize as one views this attractive town that Mark Twain in "Roughing it" could call this place in the middle sixties a decayed mining town. Jamestown also is flourishing, despite the closing of most of the near-by mines. Mariposa, at the south end of the belt, is the seat of Mariposa County.

PHYSICAL FEATURES

The Mother Lode belt is for the most part a hilly country of moderate relief. Its altitude above sea level is about 2,700 feet at the north end and 2,000 feet at Mariposa. In the most productive part of the belt, which extends from Plymouth to Jackson, in Amador County, the average altitude is about 1,500feet.

The topography is rolling except where the belt is crossed by the larger streams that flow down the western slope of the Sierra Nevada. These streams—South Fork of American River, Mokelumne, Stanislaus, Tuolumne, and Merced Rivers, to name only the master streams—cross the gold belt in deep canyons. The cutting of these canyons and the tributary

canyons that extend back from them has locally produced stretches of rugged country. The tributaries, however, head in areas of very subdued topography or even of flat country, as at Plymouth. These areas are obviously remnants of an old erosion surface that have not yet been dissected on account of their distance from the main drainage lines.

The country has in general a distinct southeast "grain," the result of the control exerted on erosion by its bedrock structure. The ridges have the same general trend and as a rule consist of resistant belts of greenstone or serpentine that have withstood erosion. They are crossed in narrow valleys by minor streams, including Amador, Sutter, and Jackson Creeks.

Some of the most rugged country in the whole belt is at the head of Moccasin Creek, a tributary that enters the Tuolumne from the south. It culminates in the peak that is well named Peñon Blanco (white cliff). Because of the broad white outcrop of a great quartz vein, which forms its crest and contrasts with its somber green cover of brush, Peñon Blanco is one of the most impressive sights on the whole gold belt. After seeing this and similar huge walls of white quartz at intervals along the gold belt, as at Carson Hill, many miles to the north, it is easy to understand how there grew up the idea of a Mother Lode, or as Borthwick wrote in 1857, in his highly interesting "Three years in California," a "great quartz vein which traverses the whole State of California."

CLIMATE AND VEGETATION

Most of the precipitation in the Mother Lode belt falls in December, January, February, and March. As may be seen from the following table, the rainfall decreases strongly southward, partly because of the decrease in altitude and partly because a southward decrease characterizes the western foothill slope of the Sierra Nevada.

Precipitation on Mother Lode belt

	Altitude (feet)	Annual precipitation (inches)			Mean annual temperature (°F.)
		Mean	Maximum	Minimum	
Georgetown	2,550	57.8	99.6 (1884)	28.1 (1898)	
Placerville	1,875	44.4	74.1 (1884)	19.4 (1888)	55
Kennedy mine	1,500	37.6	55.2 (1894)	16.7 (1908)	58.5
Sonora	1,825	36.1	59.9 (1894)	18.8 (1898)	

The following data, courteously supplied by the United States Weather Bureau from its station at the Kennedy mine near Jackson, show well the equable character of the temperature of the belt. They are based on a series of records from 1913 to 1923.

sometimes reached. South of Sonora, which is just east of the belt, the climate becomes increasingly arid, as is indicated by the character of the vegetation. Near Mariposa, as shown by the vegetation and perhaps even more impressively by the weathered

FIGURE 1.—Index map showing the location of the Mother Lode belt, California

Mean temperature (°F.) at the Kennedy mine, 1913–1923

January_____43. 6	June_____ 68. 7	November___ 51. 8
February____ 47. 6	July_____ 75. 2	December___ 44. 9
March_____ 51. 3	August_____ 74. 1	———
April_____ 54. 9	September___ 68. 5	Mean annual 58. 5
May_____ 60. 6	October_____ 60. 3	

In summer temperatures exceeding 100° are not uncommon along the Mother Lode belt, and 115° is

forms of the granodiorite, which are those of a warm arid landscape, the climate is typically arid, evaporation greatly exceeding precipitation.

The general aspect of the belt is parklike, with rounded grassy hills dotted with oaks or pines. The view northward from the Central Eureka shaft, shown in Plate 1, *A*, is typical of the Amador section. In the spring it is a region of verdant loveliness, but as the hot summer lengthens it becomes dry and parched. Soon after the first rains, which, however, may be delayed until mid-November, the grasses spring up quickly.

The vegetation is determined to some extent by the nature of the bedrock. Serpentine exercises the most notable control, areas of that rock being thickly covered with a somber-green growth of greasewood. From any vantage point, such as Carson Hill, the larger areas of serpentine can plainly be seen contrasting with their surroundings. (See pl. 1, *B*.) Chaparral becomes common south of Tuolumne River, and much of the region is covered with an impenetrable thicket of greasewood, manzanita, and scrub oak.

FIELD WORK AND ACKNOWLEDGMENTS

PREVIOUS GEOLOGIC STUDIES OF THE DISTRICT

The first study of the Mother Lode belt as a unit was made by Fairbanks[1] in 1890. Three months was spent in the field, and a belt 4 miles wide through which the lode runs was mapped and the mines were examined. As the law required that the report be completed by December of that year, only two months was available in which to study the material gathered and to write the report. A generalized geologic map on township plats accompanies the report. Because of the inadequate time in which to prepare it, the systematic and generalized portion of the report is extremely short. The rocks, the fundamental evidence on the geology of the region, could not be adequately studied, and consequently the names used have largely been superseded as a result of later work. Yet in view of the pressure under which this work was done, the report must be regarded as a notable achievement.

In the late eighties geologists of the United States Geological Survey—Becker, Turner, and Lindgren—began the detailed mapping of the lower western slope of the Sierra Nevada; and beginning with the Placerville geologic folio in 1894 there appeared a succession of folios—the "gold-belt folios"—that have been the firm foundation for all later geologic work. Professor Lindgren has recounted before the Boston Geological Society, in February, 1923, many of the difficulties with which the early workers were confronted in the perplexing rocks of the Sierra Nevada. In the preparation of these folios study of the mining geology was mainly incidental to the unraveling of the complex areal geology.

Because of the great importance of the Mother Lode belt, a special report on its mines was planned by G. F. Becker. As the necessary foundation for this report, F. L. Ransome mapped in detail a belt 70 miles long and 6.5 miles wide, extending from Plymouth to Merced River. The scale of the map— 1 mile to the

inch instead of approximately 2 miles to the inch (1:125,000)—allowed a far more detailed presentation of the topography and geology of the belt inclosing the gold-bearing quartz veins of the Mother Lode system than had been possible in the Jackson and Sonora folios. The field work, arduous physically and difficult geologically, was completed in 1898, and the results were published in the Mother Lode folio.[2] It is a pleasure to testify to the remarkable excellence and accuracy of the geologic maps in that folio, and my increasing familiarity with the region during two field seasons has steadily deepened the admiration they inspire.

Doctor Becker's studies of the Mother Lode were interrupted in 1899 by his transfer to work in the Philippines. They were never resumed, and his projected report on the Mother Lode system was not completed.

In 1900 Storms[3] published an interesting report on the Mother Lode belt, in which the main emphasis was placed on methods and costs of mining.

PRESENT INVESTIGATION

In 1915 I was assigned by F. L. Ransome, then in charge of the United States Geological Survey's metalliferous investigations, to make a detailed study of the gold-bearing quartz veins of the Mother Lode belt. Work was begun at the north end of the belt on August 15 and by the end of the field season on November 11 had been extended as far south as the South Eureka mine, near Sutter Creek. All active mines were carefully examined and mapped. During the last few weeks of this work I was ably assisted by J. Fred Hunter, whose lamentable death in 1917 cut short a life of great promise.

Field work was not resumed in the following year, on account of the lack of sufficient funds, the Secretary of the Interior believing that the Geological Survey should use its resources on the investigation of districts of which less was known than of the Mother Lode. As a consequence of the European war, the American mineral industry was stimulated to an extraordinary degree, and the demands for investigations of mineral districts by the Geological Survey were unprecedentedly numerous.

Not until 1924 was it possible to resume the investigation of the Mother Lode district. In that year the State mineralogist of California, Mr. Lloyd L. Root, offered to cooperate with the United States Geological Survey by supplying $1,000 to defray part of the expense of completing the field work. This offer was accepted, and work was resumed on June 30 and finished at Mariposa on September 20. The Plymouth and Central Eureka mines were reexamined in order to

[1] Fairbanks, H. W., Geology of the Mother Lode region: California State Min. Bur. Tenth Ann. Rept., pp. 23–90, 1890.

[2] Ransome, F. L., U. S. Geol. Survey Geol. Atlas, Mother Lode District folio (No. 63), 1900.

[3] Storms, W. H., The Mother Lode region of California: California State Min. Bur. Bull. 18, 154 pp., 1900.

view the deeper developments, but many of the mines that were operating in 1915 had shut down owing to the severe fall in the purchasing power of gold. During this field season I was efficiently assisted by Thomas B. Nolan, who later also assisted in the office during two months in compiling some of the mine descriptions that accompany this report.

Cordial acknowledgments are due to the operators along the Mother Lode belt, all of whom fully cooperated in furthering the purpose of this survey. My grateful thanks for many courtesies extended are especially due to Mr. A. S. Howe, of the Central Eureka mine; Mr. E. C. Hutchinson, of the Kennedy mine; Mr. W. J. Loring, then manager of the Plymouth and Carson Hill mines; and Mr. O. McCraney, at that time in charge of the Eagle Shawmut mine.

HISTORICAL SKETCH

The Mother Lode belt traverses a region rich in historical and literary associations. Coloma, where placer gold was first found in 1848, lies but a few miles west of the belt, and in the feverish years that followed the discovery of gold the main towns on the belt were all active centers of placer mining. Placerville, although now a tranquil fruit-shipping center, carries in its name forever a reminder of its origin. Much of the local color in Bret Harte's stories came from the Mother Lode belt, and it was at Angels Camp that Mark Twain heard the local version of the "Jumping frog of Calaveras," the story by which he was first to establish his literary fame. Some of his mining experiences on the lode are related in "Roughing it." The history of the Mother Lode belt is a great theme, but in this report only an account of its mines and of some of the ideas that have gradually developed concerning them can be sketched.

The first gold-quartz vein found in place in California was discovered at Mariposa on Colonel Frémont's Mariposa grant, near the south end of the Mother Lode belt, in August, 1849. Little more than a year had gone by since Marshall had picked up the first bit of placer gold at Coloma. Bayard Taylor,[4] who made a tour of the gold region in 1849 and wrote an interesting account of those stirring times, was told by Frémont that representative surface quartz carried 2 ounces of gold to every 25 pounds; also that the vein had been traced for a mile and was 2 feet thick at the outcrop, "gradually widening as it descends and showing larger particles of gold."

The find produced a great sensation, and Taylor concludes his account of the discovery thus prophetically, in words in which we can still feel the exuberant faith that makes mining camps:

The Sierra Nevada is pierced in every part with these priceless veins, which will produce gold for centuries after every spot of

earth from base to summit shall have been turned over and washed out.

Other quartz veins were discovered in the same year.[5] Frémont's vein, however, was the only one actually opened, and the belief that the quartz carried $2.50 a pound was proved to be too optimistic.

The first geologic examination of the Mother Lode belt was made in 1849 by Tyson,[6] a careful and acute observer. He took too seriously, perhaps, the advice he had heard that in California "it is safest to believe nothing you hear and doubt half you see." He recognized that the auriferous region was underlain chiefly by slates—clay slate, hornblende slate, chlorite slate, and talcose slate—alternating with intrusive traps or trappean rocks. Some intrusive masses of serpentine were recognized. The variety of rocks, he says, was less than he expected, but in this conclusion he was in error, for the variety is legion; however, in 1849 it was a natural error, for the science of rocks had hardly been born. Indeed, the attention of one reading the older reports on the region is constantly struck by the fact that the rock determinations of the various observers made little impression on their successors. Tyson's trappean rocks, for example, became Trask's dolerite amygdaloids and greenstone, Blake's conglomerates, and Whitney's porphyritic green slates. This state of affairs was the natural outcome of the fact that the names given to rocks were as a rule unsupported personal assertions. An authoritative interpretation of the rocks of the gold belt was not reached until careful microscopic and chemical work was done by the United States Geological Survey in the nineties.

Tyson admits that the veins are probably exceedingly numerous but on the whole he is rather skeptical that they could be worked at a profit, for he sets $20 as the minimum for payable ore. All analogy or previous experience, he says, has shown that there are few auriferous veins in the world that contain sufficient metal to pay expenses. In this cautious judgment there is doubtless a reflection from the famous dictum of Sir Roderick Murchison that deep mining for gold in the solid rock can never be profitable, because "gold was the last created metal and only occurred, therefore, in the uppermost parts of any formation."[7]

The first quartz miners were Mexicans, who used the slow but efficient arrastre to separate the gold from the ore. "As far as can be learned," says Fairbanks,[8] "the term Mother Lode was first applied to the veins worked at Nashville, 12 miles south of Placerville, Eldorado County, in the latter part of 1850 or earlier part of 1851." In the origin of this term we can undoubtedly see the influence of the early Mexican miners, for each of the great Mexican mining dis-

[4] Taylor, Bayard, Eldorado: Adventures in the path of empire, p. 110. New York, 1850.

[5] Lyman, S. C., Gold of California: Am. Jour. Sci., 2d ser., vol. 9, pp. 126-129, 1849.
[6] Tyson, P. T., Geology and industrial resources of California, 74 pp., 1851.
[7] Geikie, A., Memoir of Sir Roderick Murchison, vol. 2, p. 134, 1875.
[8] Fairbanks, H. W., Geology of the Mother Lode region: California State Min. Bur. Tenth Ann. Rept., p. 23, 1890.

tricts had its veta madre (mother lode). It is fascinating to trace the growth of the Mother Lode concept, which has become an imperishable part of the heritage of the Mother Lode counties. The immense croppings of quartz, as at Peñon Blanco and Carson Hill, viewed in the optimistic atmosphere of a mining district, irresistibly support the idea of a single great persistent quartz vein, and as early as 1857 Borthwick, who gives us accurately the local color of the times, writes of the great vein that traverses California from end to end.

Some of the first writers on the geology of the gold belt adopted whole-heartedly the Mother Lode idea, referring to the supposed continuous lode as the "Great Vein." In 1868 Browne [9] wrote, "The Mother Lode is in many respects the most remarkable metalliferous vein in the world." In 1869 Raymond,[10] his successor as United States Commissioner of Mining Statistics, felt impelled to discuss "What is the Mother Lode?" but because detailed information was wanting he was unable to reach a definite conclusion.

Gradually, however, clearer understanding came. Turner and Lindgren [11] found that the Mother Lode "must not be considered as a continuous vein, but rather as a belt of parallel though sometimes interrupted quartz-filled fissures." Storms in 1900 went so far as to say that the name Mother Lode is "unfortunate." Ransome, in particular, has shown how greatly the Mother Lode concept diverges from the actual facts and how much endless and unprofitable discussion it entails. But logical analyses of this kind, however acute and well reasoned, fail to shake the local belief.

In the early fifties a boom in quartz mining set in, despite inexperience and the crude machinery and methods available for extracting the gold. Here and at Grass Valley began the development of the California stamp battery and milling practice, which were destined to spread the world over. At first many rushed in without fully realizing the difficulties and the need of adequate capital. In 1853 there were 17 mines at work along the Mother Lode, and in 1854 the number had increased to 31. In that year the boom collapsed and left the industry under an undeserved stigma. Furthermore, a great obstacle at this time was the insecure tenure by which the mines were held, for they were subject to the government of the majority of the people of the district in which they were situated, who were placer miners.

The progress made in those early years is excellently presented in the annual reports of the first State geologist, John B. Trask. The sulphurets, as the metallic sulphides were then generally called (and are still called along the Mother Lode), were being thrown away with the tailings, and the folly of this procedure was repeatedly emphasized by Trask. He roasted the sulphurets, washed and amalgamated them, and thus showed that the sulphurets of the Eureka mine contained $130 a ton and those of the Spring Hill $270 a ton. Even as late as 1861 the pyrite in the ore of the Eureka mine, amounting to 2 or 3 per cent, was allowed to run to waste.

The Eureka and Keystone mines were the most successful during the first decade, as in fact they continued to be during many of the later decades. The Keystone paid monthly dividends of $200 a share, and in March, 1855, it paid $550 a share. The average recovered value of the Eureka ore was $20 a ton. The low wages of miners in 1855, $70 a month at the Eureka, already contrasted strongly with the fabulous wages paid to placer miners a scant three or four years earlier.

Although a few mines were highly successful during the first decade, most of them were timbered insecurely or not at all, a makeshift, unsystematic policy was followed, and the cash balance was immediately disbursed in dividends. Not many mines were likely to survive this treatment. Consequently it could be said by J. Ross Browne [12] that little progress in quartz mining had been made before 1860.

The largest operations about this time were on the Mariposa estate. Five mills were operated, with a total of 156 stamps. The largest were the two known as the Benton mills, of 64 stamps, which were built on Merced River, across from what is now Bagby station on the Yosemite Valley Railroad. They were celebrated in California, as Raymond tersely puts it, "for their cheap crushing and great loss of gold." Estimates were that the losses ran as high as 70 per cent. Although the supply of ore on the estate was regarded as inexhaustible, operations soon came to grief, partly from mismanagement (contemporaries were inclined to attribute it wholly to this cause) and partly from the exhaustion of the high-grade surface ores.

Early in the history of the Mother Lode belt Amador County established its claim as the most successful quartz-mining county. The Hayward, or Amador mine, or Old Eureka, as it has come to be known, was the most famous in those days. By 1870 it had already attained a depth of 1,350 feet and had the deepest shaft in the United States. Its ore, wrote J. Ross Browne, was mainly of low grade, "not yielding probably more than $10 to $15 per ton."

In the later sixties chlorination plants began to be built at the larger mines, and custom plants were established at certain localities, in order to extract the gold from the sulphide concentrates by means of a chloridizing roast. This mode of treating the sulphides has now been superseded, and the chlorination plants have one by one been abandoned, although the last of

9 Browne, J. R., Mineral resources of the States and Territories west of the Rocky Mountains for 1867, pp. 14–16, 1868. The section on the Mother Lode was probably written by John S. Hittell. (See p. 10.)
10 Raymond, R. W., idem for 1868, pp. 11–12, 1869.
11 Turner, H. W., and Lindgren, Waldemar, U. S. Geol. Survey Geol. Atlas, Placerville folio (No. 3), 1894.
12 Op. cit., pp. 12–13.

them, those at the Kennedy and Eagle Shawmut mines, survived until as late as 1915.

In 1872 there were in Amador County 35 quartz mills, dropping nearly 600 stamps. The stamps were light, averaging about 500 pounds, and their capacity was correspondingly small. Wages had become very low—$60 and $50 a month, with board, for first and second class miners. These low wages, which persisted with little increase until the World War, were always a prime factor in the low mining costs achieved along the Mother Lode belt. A long-forgotten episode was the employment of Chinese as miners. They proved to be hard-working, efficient miners; indeed, they shirked work far less than the white men, but as their wages were half those of white labor they were a source of friction and finally had to go.

In 1874 only two mines on the Mother Lode belt were producing annually more than $100,000—the Eureka and the Keystone—both of them in Amador County. The data in the following table [13] are therefore of considerable interest as giving a statistical picture of the two most productive mines on the lode at that time.

Production of Eureka and Keystone mines, February, 1874–March, 1875

Mine	Tons milled	Average yield to a ton	Total yield	Number of stamps
Eureka	22,098	$10.84	ᵃ$259,971	40
Keystone	25,146	18.00	452,507	40

Mine	Cost of mining per ton	Cost of milling per ton	Miner's wages per day	Weight of stamps (pounds)
Eureka	$8.03	$1.04	$3.00	------
Keystone	5.12	2.04	3.00	700

ᵃ Includes $20,254 from 219.5 tons of sulphurets at $92.27 a ton.

Raymond's report for 1875 gives a particularly complete picture of the condition of quartz mining in the Mother Lode belt. The Eureka was still the deepest mine, having attained a depth of 2,000 feet, and was the deepest mine in the United States. The next deepest was the Oneida, but it was only 1,000 feet deep. Seventeen mills were working in the Amador section and crushed in that year 85,780 tons of ore. This very moderate tonnage may be compared with the maximum output, 820,000 tons, achieved in 1915. The lowest average yield, $6 a ton, was at the Phoenix mine at Plymouth; the highest average yield, $28 a ton, was at the Original Amador. The average cost of mining and milling in Amador County was $6.50 a ton.

No marked success had so far been achieved in the mines of Calaveras County, except the Paloma, or

Gwin mine, as it came to be known later. Although quartz mining had begun early in the history of the lode, most of the mines had been forced to suspend operations. In the nineties, however, Angels Camp became a notable quartz-mining center. In fact, the Utica group of mines (Utica, Stickles, Gold Cliff, and Madison), which was under one management, ranked as the largest producer of gold bullion in the United States. In this decade the annual output of Calaveras County exceeded that of Amador.

It was considered a notable step in advance that when the Gwin mine was reopened in 1894, after an idleness of 12 years, the new shaft was sunk vertically.[14] This shaft was not, however, the first vertical shaft, for the Pacific shaft, at Plymouth, had been sunk vertically 1,600 feet some years earlier. The example now set by the Gwin was soon followed by others. The Oneida was reopened in 1896 by means of a vertical shaft, the Utica-Stickle mine sank a new vertical shaft, and the Wildman began sinking a vertical shaft 1,000 feet east of its inclined shaft. In 1898 the Kennedy began to sink its vertical shaft, which cut the vein at a depth of 3,750 feet. This shaft is now 4,500 feet deep, and ore is hoisted in it at the rate of 2,000 feet a minute. It is an extremely efficient shaft and has fully justified the wisdom of the policy adopted in sinking it. Mother Lode shafts that are sunk on the veins are crooked and expensive to maintain in alinement because of swelling ground, and they consequently allow of but moderate hoisting speeds.

There had been a steady decrease in the cost of mining and milling, due to improved powder, improved machinery, improved metallurgy, and improved mining methods. In the late nineties the Wildman mine, at Sutter Creek, a mine that represented average conditions, achieved an average cost of mining and milling of $1.97 a ton, as shown by an itemized cost sheet presented by Storms.[15]

The beginning of the new century saw the widespread adoption of electric power in place of water or steam power. This helped further to reduce costs. Under the favorable conditions then prevailing old mines could be profitably reopened.

In 1899 the Bunker Hill, near Amador City, was reopened. In 1900 the new 60-stamp mill of the Oneida was completed. The Fremont mine was reopened in this year, and two years later a 40-stamp mill was erected. At this time also the Eagle-Shawmut mine added 60 stamps to its 40, thereby increasing its milling capacity to 500 tons a day.

An event of unusual interest was the reopening of the Plymouth mine in 1911, under the leadership of W. J. Loring. This famous old mine had been highly productive in the late eighties; it had a handsome

13 Mineral resources of the States and Territories west of the Rocky Mountains for 1874, pp. 12, 74, 1875.

14 Browne, R. E., The Mother Lode of California: Min. and Sci. Press, vol. 76, p. 105, 1898.
15 Storms, W. H., The Mother Lode region of California: California State Min Bur. Bull. 18, pp. 36–37, 1900.

dividend record, and it had achieved remarkably low mining costs—less than $3 a ton. But fire and impoverishment of the ore shoot below the 1,600-foot level afflicting the mine simultaneously, it had shut down and had lain idle more than 20 years. The success that crowned the reopening and continued for several years exerted a highly stimulating influence on the mining industry by encouraging other projects for reopening idle mines.

The low-water mark for mining costs was reached in 1908–1910 by the Melones Mining Co., when the cost, including all charges except for marketing the concentrate, was $1.08 a ton. This very low cost was made possible only by the combination of the favorable physical conditions that exist at the Melones mine— a large ore body that could be worked by the glory-hole method of mining and the ore trammed out through a deep-level adit.

As late as 1914 the cost of mining at the Gold Cliff mine, at Angels Camp, a mine that required little or no timber, was kept below $2 a ton, an achievement of which its manager justifiably felt proud.

In 1916 the Eureka mine, at Sutter Creek, which had been idle for 30 years, was purchased for $500,000. It, too, had shut down as a result of fire and of diminution of its ore shoot. As years went by, generous legends began to cluster around the mine; its output became $20,000,000 or more, although a critical appraisal shows that the total can not have exceeded $12,000,000 (see p. 59), and the fire became the sole reputed reason why the mine had shut down. These legends were not without influence even on the judgment of mining engineers. Amid general public rejoicing the work of reopening the mine began. After the expenditure of another $500,000 without finding ore, work ceased in 1921. Lately, however, the Central Eureka, entering Eureka ground across its end line at greater depth, has found valuable ore, and to avoid threatened litigation bought the Eureka for $150,000.

That the Mother Lode belt might still hold some remarkable surprises, even after nearly 70 years of mining, was shown in 1919 at the Melones mine, on Carson Hill, which W. J. Loring had bought the preceding year for $600,000. The rich ore body on the hanging wall of a thick quartz vein was cut on the 300-foot level, and from this ore body was extracted within the next five years more than $5,000,000.

During the last two decades the leading mines have attained great depths, the Kennedy and Argonaut both mining ore from a vertical depth of 4,500 feet. By 1900, after 50 years of mining on the lode, a depth of only 2,250 feet had been reached in the deepest mine, the Kennedy, but in the next 25 years the depth was doubled. By some engineers familiar with operating conditions on the lode, 6,000 feet is estimated to be the limit in depth to which mining can be profitably extended.

Although the mines are favored by a low geothermal gradient, 1° F. for every 150 feet of depth, the underground working conditions, because of the high humidity—practically 100 per cent, as determined during ventilation studies by the United States Bureau of Mines—are disagreeable to native-born or European laborers. Consequently the underground forces in the deeper mines have become during the last decade almost wholly Mexican.

In looking back over the long history of the Mother Lode belt we see that few of its mines have operated continuously since the beginning of quartz mining. The Keystone, which began working in 1852, holds the record for length of continuous operation, for it was worked practically without interruption until 1920, when it shut down. Of the mines that are working to-day the Kennedy, reopened in 1885, holds the record for length of continuous operation. The history of most of the famous Mother Lode mines shows that periods of activity, often brilliant bursts of prosperity, have alternated with long periods of idleness. Equipment and machinery are removed in the idle period, and little is visible at the mine except dumps and a shaft full of water. Maps and records become scattered and lost; traditions grow up that the mine was closed because of fire, or threatened litigation, or some reason other than that the ore in sight had been exhausted and the funds necessary for exploration had already been disbursed as dividends. To reopen such a long-idle mine requires great faith and great financial courage. Great faith, because the available information concerning many of the mines is mainly that carried in some one's memory and perhaps obtained at second hand; great financial courage, for a million dollars may be risked, as in the reopening of the Plymouth mine and in the attempt to reopen the Eureka mine.

The suggestion has been made that it should be made legally obligatory to put on public record the maps and descriptions of mines about to close down, but despite the many apparent advantages of such a a requirement it is doubtful whether the plan could be enacted under our conceptions of government.

The rise in prices and wages after the World War dealt a heavy blow to the mining industry on the Mother Lode belt. Many mines were forced to shut down, and only eight were working in 1924. The possibilities of the lode are so great and its vitality is so strong, however, that we may confidently look forward to a gold production that will continue at the present rate, or even at an increased rate, for many decades to come.

OUTPUT OF GOLD

The total output of gold from the Mother Lode system to the end of 1924 is estimated by the California State Mining Bureau to be roundly $240,000,000. More than half of this sum was produced from the

10-mile section of the Mother Lode system in Amador County extending from Plymouth to Jackson. Other notably productive localities have been Angels Camp (now idle) and Carson Hill. Within the 10-mile section in Amador County are situated most of the mines having the highest individual total outputs, including the Argonaut, Eureka, Keystone, Kennedy, and Plymouth. Furthermore, all the very deep mines—those exceeding 4,500 feet in vertical depth—are in this portion of the lode. Only at the Melones mine, on Carson Hill, has a comparable depth been reached.

The output of the Mother Lode counties during recent years is shown in the subjoined table and in Figure 2, which have been compiled from the annual volumes of Mineral Resources of the United States. For further details of production these volumes can be profitably consulted. Most of the gold can be credited to the Mother Lode system, though not all, for some gold is produced by mines that are not on the Mother Lode belt itself.

Gold and silver output of Mother Lode counties, 1913–1926

| Year | Ore treated at gold milling plants (short tons) | Value of metals recovered | |
		Total	Average per ton of ore
1913	1, 242, 343	$4, 728, 450	$3. 81
1914	1, 243, 529	5, 075, 522	4. 08
1915	1, 521, 847	6, 349, 772	4. 17
1916	1, 393, 788	5, 853, 618	4. 20
1917	1, 236, 903	5, 130, 682	4. 15
1918	845, 802	4, 334, 061	4. 27
1919	780, 673	4, 894, 839	6. 27
1920	440, 516	3, 460, 423	7. 85
1921	540, 541	3, 720, 531	6. 88
1922	566, 494	3, 730, 314	6. 58
1923	402, 123	3, 142, 529	6. 77
1924	476, 949	3, 337, 949	7. 00
1925	437, 409	3, 137, 150	7. 17
1926	410, 243	3, 048, 784	7. 43

The striking features shown are the severe decline in tonnage of ore mined and the increase in the average content from $4 to $7 a ton. This increase is very closely proportional to the decrease in the purchasing power of gold, which has forced those mines that were operating on $2 to $3 ore to shut down.

Although the Mother Lode district was famous for the low-grade ore worked before the World War, yet an analysis of the statistics of output shows that even then the average value of the ores of those mines that were paying dividends was $5 a ton. It probably exceeds $8 a ton now.

The output of the deep mines of Amador County during recent years is given in the following table. As there are practically no mines in the county except those on the Mother Lode system, this table gives a faithful picture of the production of this portion of the gold belt. Amador produces two thirds of the output of the Mother Lode counties.

Deep-mine gold output of Amador County, 1913–1926

| Year | Ore treated at gold milling plants (short tons) | Yield | |
		Total	Average per ton
1913	615, 167	$2, 885, 213	$4. 64
1914	629, 037	3, 077, 940	4. 90
1915	819, 536	3, 903, 700	4. 76
1916	730, 098	3, 671, 701	5. 03
1917	798, 952	3, 667, 154	4. 59
1918	493, 584	2, 990, 195	6. 06
1919	428, 053	2, 779, 876	6. 49
1920	205, 005	1, 659, 246	8. 15
1921	279, 049	1, 959, 325	7. 02
1922	336, 564	2, 198, 542	6. 53
1923	212, 184	1, 719, 798	8. 11
1924	312, 326	2, 360, 116	7. 56
1925	281, 060	2, 176, 150	7. 74
1926	280, 502	2, 168, 969	7. 73

The individual output of the more notable mines is shown in the following table, in so far as it can be credibly ascertained. That of many important mines is missing, however, including the Kennedy and the Harvard of Jamestown. The total amount produced by the mines listed in the table is $138,000,000.

Output of Mother Lode mines to end of 1924

Argonaut	$12, 500, 000
Bunker Hill	5, 000, 000
Central Eureka	6, 000, 000
Church-Union	5, 461, 000
Dutch	2, 500, 000
Eagle-Shawmut	5, 000, 000
Eureka	12, 000, 000
Fremont and Gover	5, 500, 000
Gwin	5, 000, 000
Jumper	3, 000, 000
Keystone	17, 000, 000
Lincoln	2, 000, 000
Melones	[a] 5, 598, 310
Montezuma	1, 000, 000
Morgan	[b] 3, 000, 000
Nyman	2, 000, 000
Oneida	2, 500, 000
Pacific	1, 500, 000
Princeton	5, 000, 000
Plymouth	12, 000, 000
Rawhide	6, 000, 000
South Eureka	2, 000, 000
Utica	14, 000, 000
Wildman-Mahoney	3, 500, 000

GENERAL GEOLOGY OF THE MOTHER LODE BELT

GEOLOGY OF THE LOWER WESTERN SLOPE OF THE SIERRA NEVADA

The rocks on the lower western slope of the Sierra Nevada fall into two broadly contrasted groups. The older group, formerly termed the "Bedrock series," is of pre-Tertiary age. It is a complex of steeply dipping rocks arranged in belts that trend parallel to the course of the range and that are penetrated by many

[a] Output of Carson Hill Gold Mines (Inc.), from beginning of operations (1919) to July 31, 1924.
[b] Prior to operations of Carson Hill Gold Mines (Inc).

bodies of igneous rocks. This complex is nearly of sole interest in considering the geology of the lode mines of the Mother Lode belt. The younger group, formerly termed the "Superjacent series," is of Tertiary and Quaternary age; it comprises the gravel deposits of rivers formerly flowing on an ancient land surface that had been developed across the edges of the bedrock complex; and volcanic rocks—rhyolite, andesite, and basalt—that in places cover the gravel.

The rocks of the bedrock complex consist of alternating belts of phyllite, amphibolite schist, black slate, and volcanic greenstone. The belts range from narrow strips to zones that are miles wide. Many of them extend the whole length of the gold region, but the narrower belts crop out as long, narrow lenses; consequently the belts dovetail or interfinger to a notable extent. The rocks strike northwest and dip, with local exceptions, steeply eastward, averaging 70°.

Although the gold region lies west of the great areas of granitic rocks that make up the higher parts of the Sierra Nevada, nevertheless the continuity of the belts of schist, slate, and greenstone is interrupted by large masses of intrusive igneous rocks, chiefly peridotite and granodiorite.

The "Bedrock series" of the gold belt, exclusive of the intrusive and extrusive igneous rocks, is divided on the basis of age into two formations, the Calaveras formation and the Mariposa slate. The Calaveras formation comprises the oldest rocks; it also predominates areally. It is made up of phyllite (glossy black slate and microcrystalline schist), quartzite, and limestone. Associated and in part interbedded with the Calaveras formation are considerable masses of amphibolite schist. From fossils, found chiefly in the limestone, the Calaveras formation is known to be in part at least of Carboniferous age. The metamorphism that has affected the Calaveras rocks differs so markedly from that which has altered the Mariposa rocks that it probably took place at an earlier time and under other conditions than those that prevailed during the crustal folding in late Jurassic or early Cretaceous time. Probably at the time when the Calaveras formation was being metamorphosed, or shortly afterward, an epoch of igneous invasion set in, during which diorite was intruded. This diorite is now for the most part schistose or gneissic as a result of cataclastic deformation.

The Mariposa slate comprises the younger rocks of the bedrock complex. It consists of black slate with subordinate graywacke and conglomerate. Associated and in part interbedded with the Mariposa

formation are large amounts of greenstone, chiefly tuff and breccia. Augite is a conspicuous mineral in the greenstone, either as porphyritic crystals or as angular grains; hence the greenstone is termed descriptively an augite melaphyre. As similar greenstone is associated and in part interbedded with the Calaveras formation, much of the greenstone that is

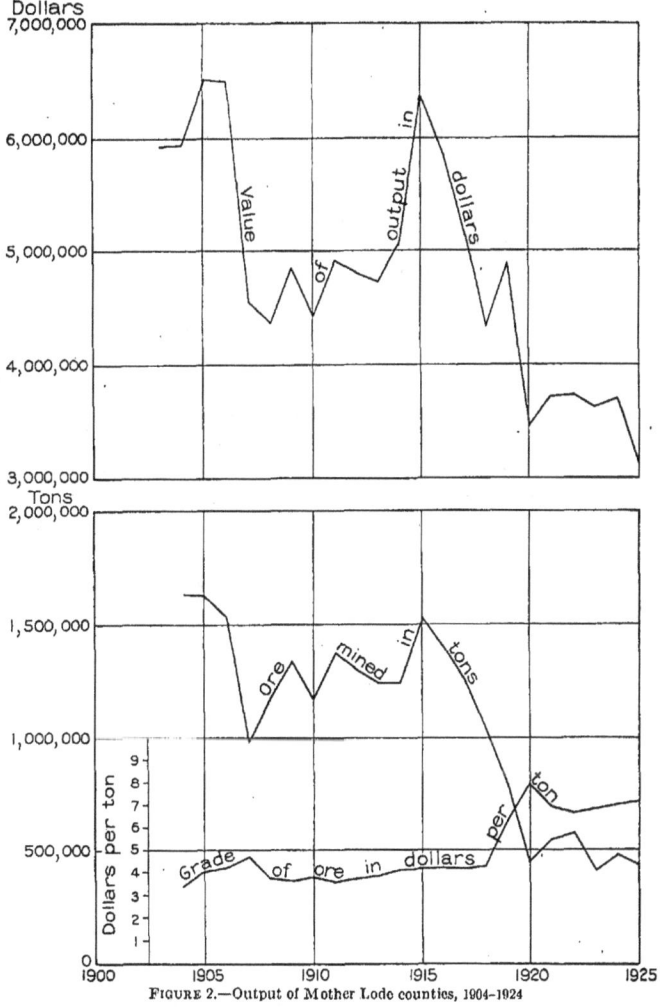

FIGURE 2.—Output of Mother Lode counties, 1904–1924

not interbedded with the Mariposa slate or with the Calaveras rocks is of undetermined age. The age of the Mariposa slate is Upper Jurassic.

At some time between the deposition of the Mariposa slate and that of the unconformably overlying Chico strata (Upper Cretaceous) the region underwent a mighty revolution known as the Cordilleran revolution. According to J. P. Smith this revolution occurred just before the end of Jurassic time. Some other geologists place it early in the Cretaceous. As a result of this revolution the Mariposa rocks were folded, closely crowded together, and complexly infolded with the Calaveras rocks. The metamorphism thereby effected, however, was of the feeblest kind. After the rocks had been folded and given their steeply dipping attitude, great volumes of magma invaded the crust. Peridotite appeared first, gabbro and hornblendite next, and granodiorite last. The granodiorite together with allied varieties—quartz monzonite, granite, and alaskite—vastly exceeds in volume the earlier rocks. The granitic rocks make up the great composite batholith of the Sierra Nevada, which forms the core of the range. Stocks and smaller batholiths of these granitic rocks appear at many places throughout the gold belt, and some occur as far west as the Great Valley of California.

During this period of igneous activity and shortly after it ore deposits of many kinds were formed. Chromite deposits were formed in the peridotites. After the granites had become emplaced contact-metamorphic copper deposits of small size were developed; contact-metamorphic tungsten deposits were formed, including the largest known representative of this group in the world; the copper-bearing lodes of the foothill copper belt, some of which are high in zinc, were formed; and, surpassing all others in economic importance, the gold deposits of the Mother Lode district and of the many other districts on the west slope of the Sierra Nevada were produced.

The distribution of these mineral deposits with reference to the intrusive rocks of the Sierra Nevada is shown on Plate 3. Probably the most striking feature shown by this map is that, as first pointed out by Von Richthofen, the interiors of the granite masses are barren of gold veins.

Since the Cordilleran revolution the gold belt has been above sea level. Only the western edge of the belt was submerged in early Upper Cretaceous (Chico) time, and for a short time near the end of the Eocene it was again submerged below the water of the sea.[16] The sediments laid down in that sea along the west flank of the Sierra Nevada compose the Ione formation, but these beds never extended as far east as the Mother Lode belt.

By the end of Eocene time the region of the gold belt had been reduced by erosion to a gently rolling landscape, diversified by longitudinal ridges of more resistant rocks that rose 1,000 to 2,000 feet above the general level.[17] In reducing the region to this condition a thickness of several thousand feet of rock was removed; and as an incident in this general reduction several thousand feet of the upward extension of the gold veins was taken off from above the present tops of the veins. The gold that was in this vanished part of the veins was concentrated in the gravel of the rivers that flowed on the old erosion surface.

In late Eocene time volcanic activity broke forth, contemporaneously with the deposition of the uppermost beds of the Ione formation, and rhyolite was erupted.[18] Later the rhyolitic outbursts were followed by immense ejections of andesite as lavas and mud flows, so voluminous that when they ceased the Sierra Nevada had been transformed into a volcanic plain. As a consequence the drainage system was thoroughly disorganized, and inevitably a new one began to establish itself. As this new drainage system developed the present landscape began to take form. The erosive power of the new streams was tremendously increased by the uplifts of the Sierra Nevada, especially by the great uplift that occurred near the beginning of Quaternary time. By this later erosion the deep transverse canyons of the present topography were cut. The significant conclusion in regard to the post-Mariposa history of the Mother Lode belt is that the topography of to-day is completely different from that which existed when the gold veins were formed.

In the following detailed presentation of the geologic features of the Mother Lode belt attention is drawn mainly to the character of the rocks that inclose the gold deposits or are closely adjacent to them. The Placerville, Jackson, and Sonora folios of the United States Geological Survey describe the geology of the region through which the Mother Lode belt extends and contain geologic maps on the scale of 1:125,000. The Mother Lode District folio gives in more detail the geology of a belt 6.5 miles wide extending from Plymouth, in Amador County, southward to Merced River, and its geologic maps on the scale of 1:63,360 are indispensable in showing the distribution of the rocks and the course of the vein system. These maps have therefore been reproduced to accompany this report as Plate 2 (in pocket).

DETAILED FEATURES OF THE GEOLOGY OF THE MOTHER LODE BELT

CALAVERAS FORMATION

The Calaveras formation, as that term is used in the gold-belt folios, includes all rocks older than the Mariposa slate, except the amphibolite schist that has been separately mapped.[19] It is considered to be mainly of Carboniferous age, but it may include some Triassic rocks and some older than Carboniferous.

As seen along the Mother Lode belt, the Calaveras formation consists chiefly of black phyllite with subordinate fine-grained quartzite, limestone, and chert. Associated and in part interbedded with the formation are green schists of contemporaneous age.

16 Dickerson, R. E., Stratigraphy and fauna of the Tejon Eocene of California: California Univ. Dept. Geology Bull., vol. 9, pp. 363–524, 1916.
17 Lindgren, Waldemar, The Tertiary gravels of the Sierra Nevada of California: U. S. Geol. Survey Prof. Paper 73, p. 198, 1911.
18 Dickerson, R. E., op. cit., pp. 409–416.
19 J. P. Smith (The geologic formations of California: California State Min. Bur. Bull. 72, p. 28, 1915) says that the Calaveras is not a true formation, but agreement with that dictum will depend on how "formation" is defined.

Carboniferous fossils have been found in some of the limestones, and the age of parts of the formation is therefore definitely known. Because of the great dearth of fossils in both the Calaveras and Mariposa formations, however, discrimination between the two must be based largely on lithology. In the Calaveras formation the argillaceous rocks are in a more advanced state of metamorphism: they are glossy slates, or phyllites, whereas those in the Mariposa formation are in the condition of roofing slates. The collateral evidence from the associated rocks supports this interpretation. For example, the quartzite of the Calaveras formation that occurs northwest of the Ford mine consists of quartz in long flattened grains, characteristic of quartz deformed under conditions of quasi-plasticity, and biotite, a mineral indicating high metamorphic intensity, occurs in sporadic minute flakes. In the Mariposa rocks, on the other hand, deformed quartz shows merely strain shadows or granulation. The associated schists in the Calaveras formation show an advanced degree of metamorphism, far higher than that shown by any rocks in the Mariposa formation, culminating in such highly metamorphic rock as the garnetiferous schist that is intercalated between the amphibolitic schists west of the Ford mine. The petrographic evidence clearly indicates that the Calaveras formation was subjected to a metamorphic action that was not experienced by the Mariposa rocks.

On account of the areal and economic importance of the green schist or amphibolite schist that is intimately associated with the Calaveras formation it is described in some detail in the following pages.

AMPHIBOLITE SCHIST ASSOCIATED WITH THE CALAVERAS FORMATION

GENERAL FEATURES

Under the symbol for amphibolite schist is shown in the Mother Lode folio a series of gray-green schists that occur as long belts and form an important element in the geology of the gold district. In Amador County they do not occur in the chief mines, but in Calaveras County, notably at Angels Camp and Carson Hill, many of the most productive mines are in this formation.

On casual view they are monotonously uniform. In many places they are strongly weathered, as a result of which they have yielded a deep-red, almost brick-red soil. They rarely show constituents that can be recognized by the unaided eye, but microscopically some are found to contain finely fibrous green amphibole that imparts its green color to the schist. Although this real amphibolite schist is possibly the main constituent of the belts mapped under the designation of this rock, yet in places the belts contain other green schists whose color is due either to chlorite or to epidote, or both. A member of the zoisite-epidote group is the most common and the most abundant constituent in the schists mapped as amphibolite schist. Furthermore the green schists contain limestone in places and are interlayered with black slate (phyllite) or with various rocks that are intermediate in composition between black slate and amphibolite schist. In other words, the name amphibolite schist has been employed as the name of a formation ("a cartographic unit"), in accordance with the not wholly defensible geologic custom of naming a heterogeneous series by a single petrographic term. This conventional geologic device has led to grave misapprehensions on the part of many who use the maps, because they expect, perhaps not unreasonably, that the maps will inform them of the precise petrographic nature of the rocks at the localities in which they are interested. If, then, the rocks in the areas mapped as amphibolite schist are found to contain other rocks, the erroneous conclusion is generally drawn that the mapping has been careless or inaccurate.

Some of the schists contain prominent crystals of hornblende, which give them a porphyritic aspect. A notable though uncommon example is the rock that occurs 300 feet east of the Gold Cliff veins at Angels Camp and is traceable for a mile. It is a roughly schistose rock, distinguished by abundant large hornblende crystals and inconspicuous feldspar phenocrysts that lie in a light-colored, nearly white groundmass. Under the microscope much granular epidote is seen, the feldspar phenocrysts prove to be albite though doubtless originally more calcic, and relics of quartz phenocrysts can be discerned. Before metamorphism the rock was probably a dacite flow, and because of its massiveness it has escaped drastic alteration.

The prominent hornblende crystals in the schists are of diverse origins: some are completely new growths—porphyroblasts produced by the conditions of metamorphism—and others are merely augite crystals that have been transformed.

In the belt of amphibolite schist just east of Plymouth several varieties can be distinguished. A schistose variolite occurs here, the only one so far found in the Sierra Nevada. A notable belt of schistose breccia, about 200 feet thick, can be traced for more than 1½ miles; it is crowded with small whitish angular fragments, which weather out in relief. Under the microscope these fragments are seen to consist of zoisite and talc, and the matrix in which they are inclosed to be made up chiefly of epidote and chlorite. Schists containing conspicuous dark hornblende crystals also occur in this belt.

Unaltered augite is common in some of the schists on the south flank of Carson Hill, and this locality has proved highly illuminating on the origin of the amphibolite schists. Augitic tuffs, augitic amphibole phyllites and schists, and amphibolite schists of various kinds occur in close proximity. Such juxtaposition of unaltered and highly metamorphic rocks such abrupt variations in degree of metamorphism, are regarded by Heim [20] as peculiarly characteristic of dynamic metamorphism. A nearly massive (non-schistose) augite basalt breccia occurs 100 feet in from the portal of the Calaveras tunnel. The fragments of volcanic rock that make up this breccia carry perfect idiomorphic phenocrysts of augite, in contrast to the broken angular particles of augite in the massive banded tuffs that are common in the Melones tunnel. Some green slate or phyllite, 1,425 feet from the portal of the Melones tunnel proves on microscopic examination to be an augitic amphibole phyllite. Highly instructive are the augitic amphibolite schists, such as the one well exposed 200 feet north of the portal of the Melones tunnel. They are greenish-gray well-foliated schists that on close examination show innumerable grains of augite. As seen microscopically the augite on the whole is remarkably intact, showing the angular outlines indicative of its pyroclastic origin. Some of the larger prisms of augite still stand athwart the foliation of the schist. The small augite fragments have been converted into actinolite fibers, and it is the parallel arrangement of these fibers that determines the foliation. In other schists conspicuous crystals of hornblende appear lying in random orientation, which are probably the larger pyroclastic augite fragments that have been paramorphically altered to hornblende, for some of them, especially those athwart the foliation, have been cracked and others sheared out into streamers; the matrix consists of amphibole, epidote, and albite. Other schists, like the zoisite-amphibole schist near the outcrop of the Stanislaus vein, are so highly metamorphic as to give no clues to their original condition.

AGE

At a number of localities the amphibolite schists, as previously pointed out, are interbedded with Calaveras black slates or phyllites and even grade into them as the result of original admixture of tuff and sediment. Such interstratification occurs at the Moore, Ford, Melones, and Eagle Shawmut mines. Between Jamestown and Stent, according to Ransome, fine-grained slaty amphibolite schists are intricately dovetailed or interleaved with the Calaveras slates. Such amphibolite schists are therefore of Calaveras age, presumably Carboniferous. Other masses are of doubtful age, and Ransome has consequently in the Mother Lode folio mapped the amphibolite schists as of "Carboniferous or Juratrias age," implying that some of them may be of Mariposa age. During the present survey no amphibolite schists were seen that are associated with rocks of established Mariposa age. The weight of evidence strongly indicates that all the amphibolite schists are of pre-Mariposa age.

STRUCTURE

The amphibolite schists have as a rule a north-westerly trend and steep northeasterly dip. The general effect is that of an isoclinal series. However, the strike of the schists varies abruptly in some localities, as at Carson Hill, and the schistosity in places diverges notably from the bedding. This evidence indicates that the beds have been closely folded. Only at the Ford mine (p. 70) were folds seen—closely appressed synclines plunging steeply northward. Major folds have not been detected.

The amphibolite schists at many places rest upon the Mariposa slate, and as in general they are older than the Mariposa rocks, such superposition indicates that it is the result either of an overturned fold or a reverse fault. Where definite evidence can be obtained, as east of Plymouth and at the Eagle Shawmut mine, the superposition of the amphibolite schists on the Mariposa beds is found to be the result of reverse faulting. Such a contact was well exposed north of Plymouth, where it had been washed bare by water escaping from a recently broken irrigation ditch; it showed that the slates are highly disturbed in a zone 2 feet wide adjacent to the contact, but the overlying amphibolite schists had been extremely crumpled and smashed through a zone 50 feet wide. These reverse faults have been the sites of subsequent quartz veining and gold deposition, as for example east of Plymouth, at Jackson Gate, and at the Eagle Shawmut mine.

MARIPOSA SLATE
GENERAL CHARACTER

The Mariposa formation consists of black slate and graywacke, with which greenstone is closely associated. Conglomerate occurs locally, and sericite schist and limestone rarely. The greenstone, because of its intimate interbedding with the normal sedimentary rocks, is in many places an inseparable part of the formation. In the gold belt folios the Mariposa formation is generally described as composed of black slate, or "clay slate, with locally varying amounts of sandstone and conglomerate." The igneous rocks, although commonly predominating in volume, are not included, and this omission illustrates the cavalier treatment that the igneous rocks formerly received in stratigraphic geology.[21]

The normal rocks of the Mariposa formation, as Ransome[22] pointed out, "constitute a remarkably persistent belt that, with the exception of a sweep to the southwestward, southwest of Fourth Crossing, which carries it for a distance outside the limits of the area mapped, traverses the district from end to end." In detail, however, this belt is locally very

[20] Heim, Albert, Geologie der Schweiz, Band 2, erste Hälfte, p. 115, 1921.

[21] Harker, Alfred, The natural history of igneous rocks, pp. 2–3, 1909.
[22] Ransome, F. L., U. S. Geol. Survey Geol. Atlas Mother Lode District folio (No. 63), p. 2, 1900.

rregular. "It branches or divides into two or more belts, and at one point, possibly at two, its continuity is interrupted." In many places along its borders the slates are intimately interbedded with augitic tuffs and breccias or interdigitate with them.

The formation trends northwest and dips as a rule steeply northeast. In places, however, it dips west, as in the stretch from Nashville to Plymouth, where the dip is 75° W. Bedding and cleavage are generally parallel, but in places they diverge. The rocks have been closely folded; in places underground the smaller or drag folds thereby produced are perfectly exposed, as may be seen to best advantage in the main adit of the Eagle Shawmut mine. (See p. 79.) The larger folds have not yet been determined.

PETROGRAPHIC FEATURES

The black Mariposa slates are fine grained and highly cleavable. Locally they have been quarried for roofing slates, as at Kelsey, a few miles north of Placerville. The few analyses available are given on page 42. The black color is produced by a carbonaceous pigment, and it is of considerable interest that this carbonaceous matter is apparently still capable of generating methane (CH_4), according to investigations of the United States Bureau of Mines. Some of the air in the Plymouth mine contained as much as 0.21 per cent of methane, which the engineers believed to be derived from the carbonaceous matter in the slate, as the methane occurred in untimbered places. In one stope it accumulated sufficiently to cause an explosion, injuring a man.

Graywacke is the name given to certain harder, coarser beds interstratified with the slates. They were called grauwacke by Fairbanks, sandstone by Ransome, and sheared sandstone by Hershey. Locally they are called diabase, diorite, and amphibolite schist; hardly ever is their true nature recognized. They are particularly useful in determining dip and strike and hence structure. As their name implies, their color is gray, and they are distinguishable from other rocks by their content of coarse grains of quartz. In addition, as found microscopically, they carry angular fragments of plagioclase, slate, quartzite, and porphyry (keratophyre?). On the one hand they grade, by diminution of grain size, into graywacke slate and slate; on the other hand, by the presence of fragmental augite, they grade into augite tuff.

They are not common in the northern part of the Mother Lode belt, where they occur merely as sporadic intercalated beds. In the southern part they are far more abundant; west of the Eagle Shawmut mine they make up a thick belt practically without any interbedded slate, and they are abundant in Hell Hollow, Merced River.

Sericite schist of light grayish-green color is interbedded and interlaminated with black slate where the Mother Lode belt crosses the South Fork of American River. Innumerable small white feldspar phenocrysts can be seen on weathered surfaces, and numerous small quartz phenocrysts appear on cross fractures. Microscopically the feldspars are found to be turbid from the multitudes of minute zoisite prisms that they inclose, but the quartz grains are nearly intact, and some still show corroded and embayed outlines. It is probable that the sericite schist is derived from dacite tuff.

Limestone as a member of the Mariposa slate was seen only near the outcrop of the Eagle Shawmut vein. A small lens of limestone was found by Turner and Ransome [23] in the Mariposa slate on Cotton Creek, in the Sonora quadrangle.

CONGLOMERATE

Conglomerate occurs in small quantities in the Mariposa slate and is of more than passing interest on account of the variety of igneous rocks in it and consequently the light that their presence sheds on the pre-Mariposa history of the region. The Plymouth mine is the only mine in which it has been encountered underground; it forms the wall rock of the vein on portions of the deeper levels.

The thickest and most persistent belt of conglomerate appears to be the one half a mile west of Plymouth; it attains a maximum thickness of 200 feet. The pebbles in it range from well rounded to subangular and are poorly shingled. Predominantly they consist of felsite, which proves to be a quartz keratophyre. Small white crystals of feldspar (albite) and quartz and hexagonal tablets of biotite are the phenocrysts, resting in a felsitic groundmass, which the microscope shows to be fluidal and locally spherulitic. The bedrock source of this quartz keratophyre is yet unknown. Other pebbles consist of quartzite, chert, quartz, aplite, and a biotitic graphophyre. The last two are particularly interesting; they are massive and undeformed and point to a period of plutonic intrusion older than that which has invaded the Mariposa rocks. The aplite consists of a fairly coarse panidiomorphic aggregate of albite, microcline, and quartz. The graphophyre resembles a medium-grained granite, but in thin sections it is seen to be a porphyry having a groundmass of well-developed graphic texture in which large plagioclase crystals are inclosed.

Coarse diorite occurs as angular pebbles in conglomerate exposed in the canyon of the Merced half a mile northwest of Kittridge.

A remarkable coarse fragmental rock, which is either a tillite or a fanglomerate, is well exposed in Dry Creek west of Drytown. The pebbles average 3 or 4 inches in diameter, but boulders as large as 3 feet occur. They range from angular to well rounded, but the poorly rounded predominate. The longer axes of flat pieces commonly stand perpendicular to the bedding. The pebbles consist of felsophyre (a devitrified keratophyre still showing perlitic cracking), quartz,

23 U. S. Geol. Survey Geol. Atlas, Sonora folio (No. 41), 1897.

quartzite, chert, and many other rocks, including limestone, mica schist, and gneiss. This tillite, as it appears most likely to be, is well exposed also on Rancheria Creek, where it shows the same features— the erratic and pockety distribution of cobbles and boulders, the slaty matrix, and the occurrence of sporadic pebbles, which though flat stand normal or highly inclined to the bedding. On Rancheria Creek the tillite is 100 feet thick and is inclosed in a belt of schistose grits, which persists southward to Amador Creek. Similar breccias occur in the Mariposa slate at Colfax, where they have been carefully studied by Moody,[24] who concluded that although they are probably tillites the evidence is not positive. Subsequently, Prof. A. C. Lawson,[25] by finding a glacially striated floor under the breccia, has proved it to be a tillite.

AGE

The Mariposa rocks within the Mother Lode belt are poorly fossiliferous. Fossils have been found at Chili Bar, on the South Fork of American River; an ammonite has been found in the slates in the Lincoln mine, in Amador County,[26] and aucellas, ammonites, and belemnites were found near Merced River on the Mariposa estate. During the present survey an ammonite inclosed in fine-grained augitic tuff was found a few hundred yards north of the Central Eureka mine in Amador County, and many aucellas were discovered in the black slates a mile southwest of Coulterville on the road to the Malvina mine. T. W. Stanton reports as follows:

> The specimens of *Aucella* from the road to the Malvina mine, 1 mile southwest of Coulterville, are identified as *Aucella erringtoni* (Gabb).
> The little ammonite from near the Central Eureka mine, half a mile south of Sutter Creek, Amador County, appears to be unlike any of the generic types that have been reported from the Mariposa. It is somewhat suggestive of *Phylloceras* but is too much distorted to permit positive classification.

The localities that have yielded the best, most distinctive collections of fossils from the Mariposa slate—the Texas ranch, southwest of Angels Camp, and Bostwicks Bar, on the Stanislaus—lie west of the Mother Lode belt proper.

The Jurassic age of the Mariposa slate was first established by means of the fossils that Clarence King found in the footwall of the Pine Tree vein, on the Mariposa estate, early in 1864.[27] This discovery was of momentous importance—in fact, it was regarded by Whitney as one of the most important achievements of the Geological Survey of California. At that time geologic ideas on the occurrence of gold were dominated by Sir Roderick Murchison's dictum

that "the Silurian and associated Paleozoic strata, together with the igneous rocks which penetrated them, have been the main recipients of gold." Consequently, the positive establishment that the rocks of the gold belt are of Mesozoic age marked a great advance in knowledge.[28]

The name Mariposa slates was given by Becker[29] in 1894 to the formation which he had earlier referred to as the Mariposa beds.[30] The age of the Mariposa rocks has been specially studied by Hyatt,[31] Diller and Stanton,[32] and Smith.[33] They agree that the Mariposa is of late, "though not latest," Jurassic age. Smith, indeed, refers it to an early Upper Jurassic age. The evidence from fossil plants harmonizes closely with this conclusion. Knowlton[34] was inclined to regard the Mariposa slate as of Middle Jurassic age.

The folding to which the Mariposa rocks were subjected and the intrusion of great volumes of granite into them, which were results of the Cordilleran revolution—the most important event in the geologic history of California, as Smith terms it—probably took place in late Jurassic time, for the revolution was of pre-Knoxville age, and the basal portion of the Knoxville itself is regarded by some geologists as of late Jurassic age.[35] "The break between Mariposa and Knoxville does not coincide with the dividing line between Jurassic and Cretaceous."[36] If these conclusions are true, then one of the mightiest of physical revolutions, one that caused the close folding of the strata of the Sierra Nevada and the subsequent intrusions of immense batholithic masses, beginning with peridotite and ending with enormous volumes of granite, ran its course during a fractional part of late Jurassic time.

GREENSTONES ASSOCIATED WITH MARIPOSA SLATE
GENERAL FEATURES

Contemporaneous volcanic rocks are intimately associated with the Mariposa slate. They are mainly the products of explosive eruptions, and the finer material is interbedded with the black slates in layers ranging in thickness from that of a sheet of paper to many yards. They occur also in long belts that are

[24] Moody, C. L., The breccias of the Mariposa formation in the vicinity of Colfax, Calif.: California Univ. Dept. Geology Bull., vol. 10, pp. 383-420, 1917.
[25] Personal communication.
[26] Turner, H. W., U. S. Geol. Survey Geol. Atlas, Jackson folio (No. 11), 1894.
[27] Whitney, J. D., Geology of California, vol. 1, pp. 482-483, 1865. Brewer, W. H., On the age of the gold-bearing rocks of the Pacific coast: Am. Jour. Sci., 2d ser., vol. 42, pp. 114-118, 1866

[28] Whitney, J. D., The auriferous gravels of the Sierra Nevada of California, pp. 34-39, 1880. Possibly Whitney overemphasizes the importance attached to Murchison's ideas. Geikie's comments on Murchison's "singularly unphilosophical notions" on the occurrence of gold are illuminating (Memoir of Sir Roderick Murchison, vol. 2, pp. 134-135, 1875).
[29] Becker, G. F., U. S. Geol. Survey Geol. Atlas, Sacramento folio (No. 5), 1894.
[30] Becker, G. F., Notes on the stratigraphy of California: U. S. Geol. Survey Bull. 19, pp. 18-20, 1885.
[31] Hyatt, Alpheus, Trias and Jura in the Western States: Geol. Soc. America Bull., vol. 5, pp. 393-434, 1894.
[32] Diller, J. S., and Stanton, T. W., The Shasta-Chico series: Geol. Soc. America Bull., vol. 5, pp. 457-459, 1894.
[33] Smith, J. P., Age of the auriferous slates of the Sierra Nevada: Geol. Soc. America Bull., vol. 5, pp. 243-258, 1894; Salient events in the geologic history of California: Science, new ser., vol. 30, pp. 347-348, 1909; The geologic formations of California: California State Min. Bur. Bull. 72, pp. 31-33, 1916.
[34] Knowlton, F. H., The Jurassic age of the "Jurassic flora of Oregon": Am. Jour. Sci., 4th ser., vol. 30, pp. 49-50, 1910.
[35] Smith, J. P., The geologic formations of California: California State Min. Bur. Bull. 72, pp. 32-33, 1916.
[36] Idem, p. 33.

free from sedimentary admixture, and some of these belts are miles wide. In these also the fragmental or pyroclastic rocks greatly predominate.

They were called by Turner and Lindgren augite porphyrite and diabase. The most common variety, according to Turner,[37] is an augite porphyrite; certainly it is the most conspicuous variety. The name augite porphyrite, highly appropriate as a descriptive term, having become obsolete in the United States, Ransome called them meta-andesites, thereby emphasizing their essentially extrusive nature. This renaming was not approved by Lindgren,[38] and in his account of the Tertiary auriferous gravel deposits of the Sierra Nevada he nowhere uses the term meta-andesite. Popularly, the terms diabase and greenstone are most widely applied to these rocks. The symbol "db," standing for diabase, was employed in many maps of the gold-belt folios to indicate the areas of "augite porphyrite and diabase," and so although no real diabase occurs in any of the mines of the Mother Lode system the term diabase is in common use and is applied to rocks that are not diabase, even under any of the various meanings of the term.[39] The well-known noncommittal name greenstone will be often used in the less precise descriptions in this report, in harmony with local usage. The blanket name greenstone is appropriate enough for those near the veins, but it is not particularly applicable to those farther away, for many of the rocks are not green but are blackish gray, bluish gray, or even lighter gray where they contain a considerable quantity of feldspar and felsite fragments. In places the name trap, applied by Tyson, the first geologist who examined them, is descriptively appropriate.

The following is a section across the greenstone belt west of Plymouth. Roughly measured from west to east it is 5,100 feet thick, it strikes N. 5° E. and dips 75°-85° E., but which is bottom or top of the section is unknown. It can be taken as representing the thicker belts of greenstone.

Section across greenstone belt west of Plymouth   Feet

1. Tuff, nonschistose, interbedded with argillite_____ 300
2. Conglomerate; contains pebbles of quartz, quartzite, chert, felsite, and hornblendic porphyry_____ 100
3. Tuff, like No. 1_____ 100
4. Fine breccias; particles do not exceed half an inch in diameter; plagioclase has been albitized and later calcitized and prehnitized; contains a sheet of augite melaphyre 110 feet thick_____ 1,450
5. Coarse augitic breccias predominant; at intervals sheets of amygdaloidal augite melaphyre; albite common, seemingly pyrogenetic but probably of metasomatic origin_____ 3,150

---

[37] Turner, H. W., Further contributions to the geology of the Sierra Nevada: U. S. Geol. Survey Seventeenth Ann. Rept., pt. 1, pp. 671-673, 1896.

[38] Lindgren, Waldemar, Econ. Geology, vol. 2, p. 87, 1907.

[39] The term diabase has recently been ruled by the committee on British petrographic nomenclature to be superfluous as a rock name (Mineralog. Mag., vol. 19, pp. 137-147, 1922).

It is bordered on the east by the conglomerate layer, 70 to 200 feet thick, described on page 13.

A marked character of the greenstones is their non-persistence and lenticularity. The lenticularity is well shown on the maps of the Mother Lode folio, where the larger masses have been delineated. The dovetailing of greenstone and slate there shown persists down to the thinnest units. On opposite sides of a crosscut, for example, the tuff beds are likely to differ in number. This nonpersistence of tuffs and breccias causes embarrassing difficulties when attempts are made to unravel the structure in a mine by sections on successive levels.

AUGITE BASALTS (BRECCIAS, TUFFS, AND LAVAS)

Augite-bearing volcanic rocks are probably the most abundant among the rocks that collectively are called greenstones. The augite makes from a third to a half of their bulk. It occurs as well-shaped crystals in the lavas and as angular grains in the pyroclastics. As a rule it is brilliantly fresh and causes the rocks to seem fresher than they are.

Massive rocks, flows or sills, occur intercalated between the tuffs and breccias but are rare. They are generally amygdaloidal. The specimen collected by Turner 2.2 miles southwest of Jackson on the road to Lancha Plana and chemically analyzed for him is representative of the massive members. Examination of the specimen, which is deposited in the United States National Museum, shows it to be markedly porphyritic, carrying many conspicuous crystals of augite in a grayish-green groundmass. Descriptively it might well be termed an augite melaphyre. It contains a few inclusions of felsite and is obscurely amygdaloidal, the amygdules consisting of chlorite, or chlorite and calcite. Under the microscope it is seen that the augite crystals are the only phenocrysts and are embedded in a groundmass composed of augite and of plagioclase microlites which are multiply twinned and too obscure to be determined. The amygdules are glass sheathed. The texture is fluidal and originally it was doubtless hyalopilitic. Chlorite is fairly common in the groundmass.

Clearly this augite melaphyre is a basalt. As its chemical analysis shows, the total iron, lime, and magnesia is higher than that of the average basalt as given by Daly.

Other analyses have not been made during the present investigation, because of the impossibility of obtaining fresh rock.

Some of the highly augitic basalts, although apparently fresh, are seen when examined microscopically to be much altered. This proved to be true, for example, of the augitic basalt that occurs on the Jackson-Sutter Creek road southwest of the Central Eureka mine; although brilliantly fresh as judged from its lustrous phenocrysts of augite, yet under the microscope it shows many feldspar phenocrysts, all of which have been wholly transformed to a white micaceous product, probably sericite.

Chemical analyses of basalts

	1	2		1	2
SiO₂	49. 24	49. 06	TiO₂	0. 96	1. 36
Al₂O₃	14. 79	15. 70	P₂O₅	. 17	. 45
Fe₂O₃	1. 36	5. 38	MnO	. 18	. 31
FeO	8. 00	6. 37	SrO	Trace.	--------
MgO	6. 89	6. 17	BaO	. 04	--------
CaO	10. 74	8. 95	Li₂O	Trace.	--------
Na₂O	2. 76	3. 11	CO₂	. 90	--------
K₂O	. 88	1. 52			
H₂O−	. 20	1. 62		100. 08	100. 00
H₂O+	2. 97				

1. Basalt near Jackson, Calif.; W. F. Hillebrand, analyst. (Turner, H. W., U. S. Geol. Survey Seventeenth Ann. Rept., pt. 1, p. 734, 1896.)
2. Average basalt. (Daly, R. A., Igneous rocks and their origin, p. 27, 1913.)

Most of the massive rocks, as already mentioned, are amygdaloidal, and neighboring amygdules are made up of calcite, chlorite, or quartz, or mixtures of these minerals. In some of the amygdaloids west of Plymouth the amygdules consist of a dark-green aphanitic substance, which under the microscope is seen to be cryptocrystalline and traversed by shrinkage cracks; hence it appears to be a metacolloid, probably an iron-rich chlorite. This same metacolloid forms amygdules in the augitic amygdaloid near the Eagle Shawmut mine, in which some of it has also developed at the expense of the augite. In this amygdaloid albite occurs in some of the amygdules— an important bit of evidence indicating that conditions were favorable for albitization.

The lavas are but faintly or slightly schistose. The tuffs and breccias are generally schistose, and at many places their schistosity or cleavage is like that of the associated slates. Differing from the similar augitic tuffs and lavas associated with the Calaveras formation, the greenstone interbedded with the Mariposa slate has not been converted into amphibolite schists. In the most advanced metamorphism found —in the belt of augite melaphyre east of the Plymouth mill—only minute fibers of actinolite have formed in the groundmass.

The intimate interbedding of the tuffs and breccias with the black slates, which has produced banded rocks of alternating green and black layers, and the common occurrence of composite rocks made up partly of argillaceous material and partly of pyroclastic matter or even intermediate between graywacke and augitic tuff point to the submarine accumulation of the tuffs and breccias and probably to their submarine ejection. The finding of an ammonite in a fine-grained augitic tuff north of the Central Eureka mine is direct proof of submarine accumulation of the tuff. In view of the probable submarine extrusion of the lavas it is surprising not to find pillow structure common. Pillow structure was found at one locality only, near the Clio mine on Tuolumne River. A belt of greenstone, 900 feet thick, rests there on Mariposa slate, which strikes N. 25° W. and dips 80° E. The basal greenstone has a remarkably fine variolitic structure through a thickness of 15 feet. It is an aphanitic green rock, crowded with varioles the size of a pea, and has an imperfect pillow or ellipsoidal structure; in the more nearly perfect ellipsoids the varioles are limited to the margins. Variolitic structure occurs elsewhere in the series at this locality. The microtexture of the heavy dark-green aphanitic greenstone above the pillow lava is unlike that seen in other greenstones of the gold belt; the augite is in long, slender prisms, many of which are markedly arcuate, with a tendency to be aggregated in radial, plumose, or "cervicorn"[40] (antler-like) growths. The feldspar also is in long, slender prisms, without fluxional arrangement.

The tuffs and breccias are highly heterogeneous in composition. They are made up of broken fragments of augite, particles of feldspar, and angular pieces of a great variety of volcanic rocks. Amygdaloid fragments are common, and neighboring amygdules have the same diversity of composition as those in the amygdaloidal lava flows.

KERATOPHYRIC ROCKS

The tuffs and breccias in the greenstones contain in places abundant white or cream-colored fragments that resemble chert. In reality the fragments are angular particles of felsite carrying sporadic small phenocrysts of feldspar and augite and prove to be keratophyre and related albitic rocks. A notable belt of such keratophyric rocks lies half a mile west of Amador City and can be traced southward at least as far as the divide between Horse and Sutter Creeks. The keratophyre fragments in the breccias of this belt range from minute particles up to angular blocks 2 feet thick. Fragmental albite also is common in these breccias and tuffs.

The keratophyres are light gray to bluish gray; they are aphanitic and fracture conchoidally. Some show fluxion structure. The phenocrysts range from a few crystals of augite to abundant crystals of plagioclase (as much as 25 per cent). Micro copically the plagioclase phenocrysts prove to be albite and resemble normal pyrogenetic crystals showing polysynthetic twinning. There is no evidence that they are the result of the replacement of more calcic plagioclase either by "autopneumatolysis" or by regional albitization, as in the Kiruna district described by Sundius.[41] However, it is highly probable that albitization of the plagioclase in the basic rocks of the greenstone series has been exceedingly common. Various

[40] Bailey, E. B., and others, Tertiary and post-Tertiary geology of Mull: Scot. and Geol. Survey Mem., p. 303, fig. 50 B, 1924.
[41] Sundius, Nils, Beiträge zur Geologie des südlichen Teils des Kirunagebiets, pp. 74-76, 1915.

kinds of albitic volcanic rocks can be distinguished— (1) a felsitic variety containing sporadic phenocrysts of augite and abundant phenocrysts of albite ($Ab_{95}An_5$) in an altered glassy groundmass containing innumerable feldspar microlites; (2) a variety of andesitic appearance, containing as much as 25 per cent of phenocrysts of albite, which are embedded in a groundmass of augite prisms and plagioclase and altered interstitial glass; (3) augite melaphyres which resemble the augite basalts but which under the microscope are seen to contain abundant phenocrysts of albite. A notable locality for such albitic augite melaphyre is the ridge west of Plymouth. The melaphyre there is an amygdaloid containing amygdules of green metacolloid; it carries prominent phenocrysts of augite and less conspicuous but more abundant phenocrysts of plagioclase. Optically the plagioclase proves to be practically pure albite, the extinction on 010 being 20° and the indices of refraction being less than that of Canada balsam. Originally the plagioclase phenocrysts were honeycombed with glass, which is now altered. Theoretically it is highly improbable that phenocrysts of pure albite should have separated from a magma so calcic as to have yielded abundant augite. Therefore, in spite of the fact that the albite crystals resemble normal pyrogenetic phenocrysts, it must be concluded that they are the result of a very perfect pseudomorphic replacement after calcic plagioclase, probably labradorite. Similar perfect albitization has affected an augite melaphyre from the 2,150-foot level of the South Eureka mine, in which many phenocrysts of augite and albite rest in a groundmass of altered glass.

CONTEMPORANEOUS INTRUSIONS

Two occurrences of intrusive igneous rocks were found that appear to be interpreted best as essentially of contemporaneous origin with the greenstones and to have been injected during the epoch of eruption.

A distinctive sheet of porphyry occurs in the greenstone belt west of Amador City and can be traced from Drytown south almost to Sutter Creek. (See fig. 3.) It is the only greenstone so far found that is of distinctive enough individuality to be traced for so long a distance. It was apparently injected as a sill; it is 500 feet thick on Amador Creek and tapers both northward and southward; on Dry Cry near its north end it is but 30 feet thick, carries inclusions of felsite,

EXPLANATION

FIGURE 3.—Geologic map of Mother Lode belt from Plymouth to Jackson

and is amygdaloidal. The striking feature of the rock is that it contains abundant large phenocrysts of tabular plagioclase in an aphanitic gray-green groundmass. The plagioclase crystals are an inch in diameter and a quarter of an inch thick; most of them are now so altered as to be beyond determination, but they appear to have been albitized and later sericitized or kaolinized. With them are associated much less conspicuous phenocrysts of augite. The groundmass was doleritic or intersertal but is greatly obscured by alteration products.

Diabase porphyry forms two dikelike masses in the Mariposa slate southeast of the Plymouth mine. It is a dark heavy rock containing inconspicuous phenocrysts of tabular plagioclase. Microscopically the plagioclase proves to be albite and is seen to rest in a diabase groundmass of plagioclase and augite. The texture of the rock is clearly that of a diabase porphyry, so that it proved highly surprising to find the feldspar to be albite; doubtless this albite is another result of the widespread albitization. At some time after the albitization the rock was partly crushed, and the albite has been partly replaced by a biotite-like mineral (stilpnomelane?).

CONCLUSION

In conclusion, the "greenstones" of the Mother Lode comprise augite basalts; albitic augite melaphyres, in which the albite phenocrysts are of replacement origin; andesites, also with albitized phenocrysts; and keratophyres, in which the albite phenocrysts are presumably primary. Means by which primary albite phenocrysts can be distinguished from those of metasomatic origin in the rocks of the gold belt are not yet known. Tuffs and breccias greatly predominate. Such predominance of pyroclastic members in a basaltic series is abnormal, but it seems more comprehensible in view of the keratophyric eruptions that contributed to the building up of the series.

INTRUSIVE IGNEOUS ROCKS
GENERAL RELATIONS

The Calaveras and Mariposa formations have been intruded by igneous rocks of many varieties. Some of these were probably intruded at the end of the period when the Calaveras rocks were metamorphosed, possibly at the end of the Carboniferous. Metadiorites, as they were called by Ransome, appear to be the main representatives of this epoch of igneous activity.

The greater number, or at any rate those occurring in immensely greater volume, are those of post-Mariposa age, which invaded the rocks of the gold belt as the final act in the great revolution that occurred at or near the end of Jurassic time. The igneous rocks of this epoch appear to have been intruded in the normal order of decreasing basicity. The oldest of the post-Mariposa igneous rocks were peridotites, which soon after their intrusion and consolidation were transformed into serpentines. Because of their prevalence within the Mother Lode belt the serpentines are of main economic interest. Hornblendite and gabbro occur only as minor masses along the gold belt. The granitic rocks that higher on the slope of the Sierra Nevada occur as vast masses of granodiorite, quartz monzonite, granite, and alaskite are represented only by the granodiorite south of Mariposa and by the quartz diorite porphyry south of Placerville, which is a quickly chilled offshoot from the granodiorite intrusion of Coloma.

Albitite and albitite porphyry dikes occur in Tuolumne County, presumably among the last igneous rocks to be injected. They are of peculiar interest because of their unusual composition and because they have been locally mineralized. They have been supposed to resemble closely the auriferous mineralized dikes of the famous Treadwell mines in Alaska, but the resemblance is much less than has been thought.[42]

METADIORITE

Under the designation metadiorite Ransome mapped intrusive masses of foliated "diorites, largely quartz diorites." They lie just east of the Mother Lode system, among them being the mass east of Plymouth, that at Mokelumne Hill, and that east of San Andreas. They differ enormously in petrographic habit, comprising types as diverse as muscovite granite and hornblendite. Manifestly the name metadiorite was employed as a convenient field term to designate the average rock. The metadiorites were regarded by Ransome as of "Jura-Trias or early Cretaceous" age. The hypothesis will here be presented that they are in large part at least far older and were intruded at the end of Carboniferous time. Probably the metadiorite is of two ages, the earlier being of pre-Mariposa age and the later of post-Mariposa age.

The metadiorite east of Plymouth, which lies nearer to a productive mine of the Mother Lode belt than any other mass of metadiorite, is highly schistose along its contact with the amphibolite schist west of it; in fact, it is difficult to determine where the diorite ends and the schist begins. The metadiorite is a light-colored foliated hornblendic quartz diorite. Unlike most of the metadiorite masses it is notably homogeneous in composition. It becomes less schistose away from the contact, in places being a fine-grained augen gneiss.

The metadiorite at Plymouth is poorer in biotite and hornblende than the granodiorite at Mariposa, which terminates the Mother Lode belt, and as it is markedly foliated, it differs greatly from the granodiorite, which is a completely undeformed rock. This strong difference suggests, though it does not prove, that the metadiorite is older than the granodiorite and belongs to an older epoch of intrusion.

The large mass of metadiorite at Mokelumne Hill is extraordinarily heterogeneous in grain and composition; prevailingly it is a basic diorite rich in hornblende; locally it is a hornblendite. It is thoroughly

[42] Knopf, Adolph, The Eagle River region, southeastern Alaska: U. S. Geol. Survey Bull. 502, pp. 28–29, 1912.

weathered to a dark-red soil 4 feet deep, in spite of the steep slopes and youthfulness of Mokelumne Canyon. The metadiorite is excellently and continuously exposed along Mokelumne River; it is cut by dikes of aplite, and these have been rendered schistose.

The large mass of metadiorite east and southeast of San Andreas is of particular interest. A roughly foliated or schistose quartz monzonite is the most common facies. As proved by the microscope, the foliated structure is a cataclastic effect, and the rock is seen to consist of microcline, plagioclase (now largely zoisite and muscovite), granulated quartz, and biotite, with accessory allanite.

At the border of the igneous mass is a highly rotten, highly schistose, highly biotitic diorite; in this border zone are sporadic lenses of the fresh, less schistose rock, which is well exposed in the road cuttings along the Willow Creek road. It is found to be composed of biotite, hornblende, quartz, and an extraordinary abundance of clinozoisite, undoubtedly derived from a preexistent plagioclase. The most interesting feature is that the biotite shows intense black pleochroic haloes surrounding minute nuclei of zircon (?). These haloes are fully developed from periphery to center and are of an intensity so far known only in pre-Triassic rocks. Examination of a large number of slides of the post-Mariposa granites of the Sierra Nevada fails to show anything but faintly developed pleochroic haloes in their biotite. Fairly strong pleochroic haloes surround also the inclusions of apatite in the biotite of the diorite of the San Andreas area.

A coarse white granite spangled with flakes of muscovite occurs near the border of the metadiorite half a mile north of Old Gulch. It is roughly schistose, is composed chiefly of microcline, quartz, and muscovite, and shows abundant evidence of cataclastic deformation. The muscovite shows some faint yellow pleochroic haloes that surround minute colorless prisms of unknown composition. The work of Mügge [43] on the experimental production of pleochroic haloes has shown that muscovite is extremely insensitive to the action of α-particles, no results being produced after four months' intense bombardment; and it is therefore probable that this muscovite in the granite southeast of San Andreas, which shows perceptible haloes as the result of the extremely feeble rate of bombardment from the radioactive inclusions in it, is of high geologic antiquity, probably pre-Triassic.

Associated with the metadiorite are dikes of aplite and hornblende lamprophyre. The aplite is a pure-white sugar-grained rock, which has been sufficiently sheared to be roughly schistose. The hornblende lamprophyre (spessartite) occurs at Old Gulch as dikes a foot or so thick cutting the limestone, quartzite, and schist of the Calaveras formation. It is a dark-gray porphyry carrying conspicuous phenocrysts of

hornblende. Microscopically hornblende is seen to predominate strongly, the interstitial material being largely zoisite. Apatite is an abundant accessory mineral. One of the dikes cuts Calaveras schist diagonally and has been sliced into segments along the foliation planes of the schist. This lamprophyre when examined microscopically shows well the early stages of mineral reconstitution due to dynamic metamorphism. Zoisite and talc are the new minerals formed.

The magma intruded east of San Andreas was evidently capable of much differentiation; it produced various quartz dioritic facies and muscovite granite, and the major intrusions were followed by a series of diaschistic dikes, consisting of aplite and hornblende lamprophyre.

The age of the epoch of intrusion of the metadiorite east of San Andreas is post-Calaveras, as the rocks intruded are of Calaveras age. That it is pre-Mariposa and probably late Carboniferous is suggested (1) by the occurrence of plutonic igneous rocks in the conglomerates of the Mariposa formation; (2) by the strong dynamic deformation that has affected the metadiorites, so unlike the granodiorite of known post-Mariposa age; (3) by the fact that the Calaveras rocks have undergone a relatively deep-seated metamorphism in contrast with the Mariposa rocks, and this metamorphism might well have culminated in an epoch of intrusion; and (4) by the deep-colored radiohaloes, which point to a geologic age much greater than that of the post-Mariposa granodiorite.

Probably the great crustal compression that folded the Mariposa slate at or near the end of Jurassic time dynamically deformed the metadiorites and their associated aplite and lamprophyre.

IGNEOUS ROCKS OF POST-MARIPOSA AGE

SERPENTINE

Serpentine occurs in long lenses and belts that have been derived from intrusive masses of peridotite. The alteration to serpentine has generally been so thorough that little remains of the original peridotite.

In Amador County serpentine does not occur near the Mother Lode system and consequently does not appear in any of the mines. In Calaveras, Tuolumne, and Mariposa Counties it occurs in great volume along the belt. The largest mass is that which extends southward from Tuolumne River along Moccasin Creek, forming the rough country at the head of that creek, culminating in Peñon Blanco Point, and narrowing down at Coulterville. South of Coulterville it again expands, forming the exceedingly rugged country of the canyon of Merced River.

The serpentine areas support a thick growth of somber green brush, which sets them off conspicuously from the rest of the country. A peculiar feature of the serpentine masses is their water-yielding capacity, which was notably shown in 1924, one of the driest

[43] Mügge, O., Radioaktivität und pleochroitische Höfe: Centralbl. Mineralogie, 1909, p. 69.

years in many decades, by the fact that though all other springs had become dry those in serpentine were still running. This capacity is evidently the result of the abundance of slickensides and shear planes in the serpentine.

Two markedly different varieties of serpentine occur. The more prevalent variety is a dark-green rock that contains numerous plates of bastite of almost metallic luster, which are scattered conspicuously through a dark aphanitic matrix; thus this serpentine has seemingly a porphyritic texture. The other variety is of obviously crystalline texture; it is a sparkling, fine-grained, homogeneous rock of paler-green color than the other serpentine. Moreover, it is massive, without the countless slickensided surfaces and shear planes of the other variety. Under the microscope this serpentine is seen to be composed of antigorite in well-defined plates, which accounts for the phanero-crystalline character of the rock. No traces of the mineral from which the antigorite was derived remain.

Antigorite serpentine occurs near the Ford mine, San Andreas, on the ridge forming the western part of Angels Camp, and at Carson Hill.

Many of the serpentine masses grade at their borders into talc schist. Evidently there has been some movement along the contacts since the intrusion of the original peridotite, and this movement has metamorphosed the peridotite or its serpentinized equivalent into talc schist.

The peridotites from which the serpentines were derived are among the oldest, if not the oldest, of the plutonic intrusions that invaded the region at or near the end of Jurassic time. They appear to represent the "basic prelude" of the tremendous igneous activity that began then. At many places they are cut by small intrusions of hornblendite and gabbro. Ransome found some conflict in the evidence on the relative age of the serpentine and the metadiorite. In some places serpentine is apparently intrusive into rather basic metadiorites, and in others dikelike masses of similar metadiorite traverse the serpentine. Probably metadiorite was formed in two periods, not one, as Ransome thought, and the earlier was of pre-Mariposa age and the later possibly of post-Mariposa age.

Although it is commonly assumed that the serpentine is all of the same age, it is also possible that some of the serpentine—that wholly inclosed in Calaveras rocks—is older than that which is intrusive into the Mariposa slate.

The date at which the serpentinization took place was determined by an unexpected bit of evidence. Antigorite serpentine from the 1,350-foot level of the Melones mine shows under the microscope that the antigorite is in all stages of invasion and replacement by sericite and ankerite. Because of the resemblance of sericite to talc the identification of the sericite was verified by obtaining the potassium flame with a hydrofluoric acid solution of some of the sericitized serpen-

tine. Here is conclusive proof that the serpentine had already been formed by the time of the advent of the ore-depositing solutions. The work of these solutions marked the end stage in the Jurasside igneous activity, and the time of serpentinization is thus fixed as having occurred within the epoch of igneous activity and post-intrusive processes. At that time the serpentine, now at a vertical depth of 1,200 feet, had an additional cover of several thousand feet of rock.

A characteristic feature of the serpentine is that at many places along the Mother Lode belt it has been altered into immense bodies of mariposite-ankerite rock laced with quartz veins and stringers. The accessory chromite in the serpentine has evidently been necessary to furnish the chromium that gives the green color to the mariposite, which is the most distinctive mineral in the alteration product.

HORNBLENDITE AND GABBRO

Small masses and dikes of hornblendite and gabbro occur throughout the gold belt. They appear to have a predilection, as it were, for the serpentine areas. Gradation between serpentine and gabbro is suggested in places, as on the brush-covered slopes of the southeast flank of Peñon Blanco, but these appearances are deceptive, for in tunnels driven under such localities the serpentine and gabbro are seen to be separate masses. Wherever positive evidence can be obtained, the hornblendite and gabbro are found to be intrusive into the serpentine.

Feldspathic hornblendite masses are especially common along Moccasin Creek; they are of many abruptly differing facies. Near Tuolumne River are particularly fine representatives. This hornblendite is made up of brilliant columns of hornblende three-fourths of an inch long, with a little feldspar filling the angular interstices between the columns. The feldspar proves under the microscope to be albite, forming large homogeneous areas that contain abundant zoisite. Evidently the albite has been derived from the break-up of a calcic plagioclase.

A small mass of saussurite gabbro occurs on the west slope of Carson Hill, intrusive into the Calaveras formation. It is a rather coarse grained heavy rock containing abundant diallage, and microscopically its original feldspar is found to have altered to an aggregate of albite and zoisite. On account of its metamorphic condition, this gabbro might be taken to be of pre-Mariposa age, but elsewhere in the Sierra Nevada saussurite gabbro is known to be intrusive into the Mariposa slate.[44]

A sheared hornblendite, which has become somewhat talcose as a result of the shearing, occurs in the Calaveras formation at the Ford mine, east of San Andreas. Microscopically it is found that some of

44 Knopf, Adolph; Notes on the foothill copper belt of the Sierra Nevada: California Univ. Dept. Geology Bull., vol. 4, p. 419, 1906. Reid; J. A., Ore deposits of Copperopolis, Calaveras County, Calif.: Econ. Geology, vol. 2, p. 391, 1907.

the deep-colored pyrogenic hornblende has been altered to pale fibrous amphibole. Titanite is a common accessory and has produced radiohaloes in the hornblende; but as hornblende has been experimentally shown to be highly sensitive to the action of α-particles, no conclusions can be drawn as to the age of the hornblendite.

Hornblendite is known to be of two ages in the Sierra Nevada: fragments of hornblendite occur in volcanic breccias associated with the Mariposa slate, and dikes of hornblendite cut the Mariposa slate, as at Copperopolis.[45]

GRANODIORITE

The Mother Lode belt, as has long been known, is terminated on the south by granodiorite that extends westward from the enormous masses of the high Sierra. The granodiorite appears at Mormon Bar, 2 miles south of Mariposa, and its contact continues westward approximately along the State Highway. It has exerted a strong contact metamorphism on the Mariposa slate, the influence of which extends as much as a mile from the contact.

The granodiorite yields immense outcrops of fresh rock, which present the forms characteristic of granite weathering in a warm arid climate, such as huge boulders resting insecurely on rock pedestals. The feldspars appear to be brilliantly fresh, but the hornblende weathers to yield a red iron pigment.

The granodiorite is uniform in texture and composition, and it contains the basic clots, both the ellipsoidal and the angular, that are so common in it in the high Sierra. The granodiorite contains roughly 20 per cent of hornblende and biotite; andesine ($Ab_{63}An_{37}$) is the predominant light-colored mineral, orthoclase is subordinate, and quartz showing strain shadows is fairly abundant. A little myrmekite occurs, and titanite, magnetite, apatite, and zircon are accessory minerals. Some of the zircons that are inclosed in biotite have developed weak radiohaloes; a zircon inclosed in hornblende has produced a fairly intense halo.

Quartz diorite porphyry is intrusive into the Mother Lode belt near the Church Union mine, south of Placerville. It is a highly porphyritic rock crowded with phenocrysts of plagioclase ($Ab_{45}An_{55}$) and quartz, which are inclosed in a sparse aphanitic groundmass. It grades into the granodiorite mass of Coloma.[46] It has exerted no perceptible contact metamorphism on the Mariposa slate into which it is intrusive.

These facts—the lack of contact metamorphism and the chilling of thick offshoots from the granodiorite magma so that the material consolidated to a porphyry having an aphanitic groundmass— show that the granodiorite magma in this part of

the gold belt ascended to a high level in the earth's crust, where the rocks were relatively cold, and they have an important bearing on our hypothesis of the origin of the gold-quartz veins.

ALBITITE AND ALBITITE PORPHYRY

General features.—Dikes and small intrusive masses of a white rock composed almost wholly of albite occur in the southern portion of the Mother Lode belt. The chief localities are southeast of Jacksonville and along the east side of the valley of Moccasin Creek. Both a granular variety resembling aplite and a distinctly porphyritic variety occur, and in accordance with the suggestion of Turner,[47] who first recognized the nature of these intrusions, they will be termed albitite and albitite porphyry respectively.

In places the dikes carry a little gold as the result of subsequent mineralization, and attempts have been made to mine the more highly auriferous portions, as at the Willieta mine, near Jacksonville, but these attempts have not met with success. That there is any genetic connection between these dikes and the occurrence of gold along the Mother Lode is not probable, for that portion of the lode along which they are most common—from Jacksonville south along Moccasin Creek—has been notoriously poor in gold.

Petrography.—These highly albitic igneous rocks were called soda syenite and soda syenite porphyry by Turner.[48] They were mapped and described by Ransome in the Mother Lode folio as soda syenite granophyre. The term granophyre was evidently used by Ransome in its original signification of a porphyry whose groundmass is holocrystalline, but inasmuch as both evenly granular and porphyritic varieties of albite rocks occur in the Mother Lode belt, it seems preferable to use the terms albitite and albitite porphyry.

The small intrusive mass of albitite that forms the ridge top north of the Clio mine consists of a white granular rock containing innumerable radial groups of a blue fibrous mineral. Under the microscope the rock is seen to be composed of an allotriomorphic granular assemblage of very pure albite ($Ab_{97}An_3$), and the blue mineral proves to be the amphibole riebeckite in delicately fibrous aggregates. Titanite and apatite are accessory minerals.

At the Willieta mine, now abandoned, there is a dike of albitite porphyry 150 feet thick. It is cut by very widely spaced quartz veinlets, 1, 2, or 3 inches thick. The unaltered porphyry shows abundant phenocrysts of albite, 4 to 5 millimeters in diameter, embedded in a microgranular groundmass. Much of the rock contains an unusually large proportion of riebeckite.

[45] Knopf, Adolph, op. cit., pp. 390–391.
[46] Turner, H. W., and Lindgren, Waldemar, U. S. Geol. Survey Geol. Atlas, Placerville folio (No. 3), 1894. It is there mapped as "quartz porphyrite."

[47] Turner, H. W., Notice of some syenitic rocks from California: Am. Geologist, vol. 17, pp. 379–386, 1896; Further contributions to the geology of the Sierra Nevada: U. S. Geol. Survey Seventeenth Ann. Rept., pt. 1, p. 665, 1896.
[48] See especially his paper on The nomenclature of the feldspathic granolites: Jour. Geology, vol. 8, pp. 105–108, 1900.

The most persistent mass of the albitic rocks is the long dike east of Moccasin Creek. This dike was thought by Ransome to show remarkable variations in texture and composition, ranging from a muscovite soda granite through soda syenite granophyre to soda granite porphyry. But the exposures are poor, as Ransome himself points out, and it is more likely that the supposed granitic facies are in reality separate intrusions of different rocks. An unimpeachably good section across the dike is exposed in a cut on the recently built Hetch Hetchy Railroad. The dike is here 25 feet thick; it dips 60° E. and is a multiple dike, in the sense that it is the result of the successive injections of magma of one kind into the same fissure. The footwall portion consists of albitite porphyry 6 feet thick; the remainder consists of evenly granular albitite. A short distance in the hanging wall is an attendant albitite dike 6 to 12 inches thick, but in spite of its thinness it is nonporphyritic, at least to the unaided eye.

The albitite porphyry contains numerous phenocrysts of albite in a gray-white groundmass that is faintly speckled with aegirine. The groundmass is microgranular, there being a marked contrast in size between the albite ($Ab_{95}An_5$) of the phenocrysts and that of the groundmass. Aegirine constitutes about 10 per cent of the rock; it is in prisms with ragged ends, and, as is characteristic of aegirine, much of it is of deuteric aspect.

The albitite, constituting four-fifths of the thickness of the dike, is faintly banded and chilled against the porphyry. It therefore represents a second injection of magma in the same fissure. It is white, even grained, and finely phanerocrystalline. It is an allotriomorphic granular aggregate of albite ($Ab_{95}An_5$), with which is associated a small quantity of aegirine, much smaller than that in the albitite porphyry portion of the dike. The features shown in this cross section of dike in the railroad cut explain some of the puzzling differences in texture noted by Ransome; but the supposed variations in composition, in view of the remarkably constant composition of the dike as it appeared to me, must be due to the doubtful interpretation of the naturally poor exposures.

The soda syenite porphyry (P. R. C. 773) analyzed for Turner [49] was obtained from the dike east of Moccasin Creek, not far from the railroad cut described in the preceding paragraph. It resembles a fine-grained aplite but under the microscope is seen to contain minute phenocrysts of albite in a groundmass of albite and aegirine. It can be termed an aegirine albitite porphyry.

Analysis of aegirine albitite porphyry from Moccasin Creek, Calif.
[H. N. Stokes, analyst]

SiO_2	67.53	K_2O	0.10
Al_2O_3	18.57	H_2O-	.15
Fe_2O_3	1.13	H_2O+	.31
FeO	.08	P_2O_5	.11
MgO	.24	TiO_2	.07
CaO	.55		
Na_2O	11.50		100.34

Contains traces of MnO, SrO, and F; BaO and Li_2O are absent.

The chemical analysis has been recast into mineral composition as follows, and the result harmonizes fairly well with the findings under the microscope. CaO is slightly low, for it is insufficient to make a plagioclase of the composition determined by the microscope—$Ab_{95}An_5 + Ab_{97}An_3$. MgO is probably low, being insufficient to make an aegirine of the composition indicated microscopically.

Albite	92.75
Orthoclase	.56
Aegirine (acmite, 3.70; diopside, 1.30; hedenbergite, 0.25)	5.25
Apatite	.27
Titanite	.20
	99.03

GEOTHERMAL GRADIENT

The rate of increase of rock temperature downward along the Mother Lode in Amador County is 1° F. for every 150 feet of depth. The geothermal gradient was determined at the Plymouth, Central Eureka, and Kennedy mines. In all determinations of rock temperature a pair of maximum thermometers was used. These thermometers, each inclosed in a steel tube, were inserted in 6-foot holes specially drilled for the purpose, in drifts recently driven. The holes were then sealed, and the thermometers were left for six hours or more, in some measurements as long as twelve hours. In some of the drill holes the thermometers were left while a full round of holes was fired in the advancing drift, 10 feet ahead. The thermometers remained undamaged. A particular difficulty that the use of maximum thermometers presented was that the rock temperature in some of the holes was less than the air temperature of the drifts. This difference made it necessary to immerse the thermometers in iced water before inserting them in the holes. It also tended to make the readings of the rock temperature slightly higher than the true temperature, as the thermometers on removal from the holes immediately began warming to the temperature of the drifts.

The rocks in which the temperature was measured are the black slates of the Mariposa formation. Some of the holes were drilled dry, but most of them were drilled wet. The greatest vertical range for which

---

[49] Op. cit., pp. 727–728; also U. S. Geol. Survey Bull. 591, p. 185, 1915.

the geothermal gradient was measured is 4,200 feet. The detailed data are given in the following table. The average gradient is 1° F. for 153 feet in depth, but 1° F. for 150 feet is believed to express the accuracy attained.

Rock temperature along the Mother Lode, California

Mine	Level	Vertical depth (feet)	Vertical difference used in determining gradient (feet)	Air temperature of drift (°F.)	Rock temperature (°F.)	Geothermal gradient (feet per °F.)
Plymouth	1,600	1,575	--------	74	72	------
Do	4,000	3,600	2,025	83	86	145
Central Eureka [a]	4,400	4,095	4,095	86	[b]83.6	160
Kennedy [c]	4,200	4,200	4,200	85	86	153

[a] The annual average temperature at the Central Eureka mine is taken to be 58°, slightly less than that at the Kennedy.
[b] Average of three pairs of readings: 82.8°, 84.8°, 84°.
[c] The annual average temperature at the Kennedy mine for the period 1913-1923, from readings taken at the mine for the Weather Bureau, is 58.5°.

For comparison the rate of increase of temperature in depth in the Grass Valley district, as found by Lindgren,[50] is 1° F. for 122 feet, measured in a vertical range of 1,523 feet.

GOLD ORE BODIES

GENERAL FEATURES

The Mother Lode belt has the general northwesterly trend of the bedrock formations, but it is linearly more persistent than the long, flat interfingering lenses in which these formations crop out. Consequently it traverses rocks of both Mariposa and Calaveras ages and the serpentine masses that intrude them. In Eldorado and Amador Counties it is inclosed in the greenstones and the black Mariposa slate; in Calaveras County south of the Gwin mine it enters the phyllites of the Calaveras formation and the green schists, within which it continues through Tuolumne County, except where it traverses certain large masses of intrusive serpentine; and in Mariposa County it returns to the Mariposa slate.

The Mother Lode belt lies considerably west of the great granitic masses of the higher Sierra. The roughly foliated variety of granitic rock termed by Ransome metadiorite locally occurs near the belt, as at Plymouth, where it crops out a few thousand feet east of the Plymouth mine, and at San Andreas.

Gold-bearing veins occur east of the Mother Lode belt, generally in granodiorite or in the bordering schists of Calaveras age. They are smaller and likely to be of higher grade. Many of the ores are high in sulphides, and pyrrhotite, a sulphide rarely occurring in the Mother Lode system, is peculiarly characteristic. A good description of these veins is given by Storms.[51]

[50] Lindgren, Waldemar, The gold-quartz veins of Nevada City and Grass Valley districts, California: U. S. Geol. Survey Seventeenth Ann. Rept., pt. 2, pp. 170-171, 1896.
[51] Storms, W. H., The Mother Lode region of California: California State Min. Bur. Bull. 18, pp. 21-26, 1900.

The gold deposits of the Mother Lode belt form two main groups—quartz veins and ore bodies of mineralized country rock. The second group is by far the more diversified, as it includes auriferous greenstone ("gray ore") and auriferous schists. Moreover, the ore bodies of auriferous schists fall again into two kinds, between which, however, there are no hard and fast lines—stockworks, which consist of masses of schist laced with gold-bearing veinlets; and mineralized schists, which are ore mainly because they have been replaced by ankerite and gold was introduced during the replacement. As the schists thus mineralized were of several kinds, the diversity of the resulting gold deposits was further increased.

Although the quartz veins that are inclosed in the Mariposa slate are much alike from mine to mine, especially in Amador County, yet the variety of ore bodies along the Mother Lode belt and the complexity of the occurrence of the ore bodies at many of the mines is far greater than earlier accounts of the Mother Lode belt would lead one to expect. For illustrations of this complexity the special descriptions in this report may be consulted, notably those of Carson Hill and of the Eagle Shawmut mine.

Ore bodies of many of the various kinds occur all along the Mother Lode belt. Certain regional differences, however, can be discerned: productive quartz veins are more characteristic of the northern portion of the lode, especially in Amador County; and the southern portion, say from Angels Camp southward, has its own particular stamp, due to the prevalence of ore bodies of mineralized country rock, immense persistent barren quartz veins, and enormous bodies of mariposite-ankerite rock.

The most productive portion of the belt is the 10-mile stretch between Plymouth and Jackson. It has produced more than $100,000,000 and is now yielding the bulk of the annual output. The depth reached here is also greatest, two of the mines, the Kennedy and Argonaut, obtaining their ore from vertical depths of more than 4,500 feet.

In the quartz veins quartz is of course the dominant mineral, ankerite and albite occurring very subordinately. Sulphides average between 1 and 2 per cent of the ore; pyrite predominates, as a rule overwhelmingly; arsenopyrite is next in abundance; followed by sphalerite, galena, chalcopyrite, and tetrahedrite. The telluride petzite is rare.

In the ore bodies of mineralized country rock ankerite is the dominant mineral, followed by sericite and by minor albite and quartz. The same sulphides that occur in the quartz veins are present and are more abundant. Pyrite and arsenopyrite are the only ones that have formed by replacement, however, the other sulphides being restricted to veinlets. Molybdenite exceptionally has formed in some quantity by replacement at Carson Hill, one of the few localities on the Mother Lode belt where it has so far been found.

The ores are of low or moderate grade; in recent years they have averaged $7 a ton. The very low grade ores, between $2 and $3 a ton, that were formerly mined no longer pay to work.

The gold bullion ranges in fineness from 790 to 840 parts per thousand, the remainder being chiefly silver. In any given mine the fineness is fairly constant; at the Central Eureka, which is now drawing its ore from a depth of more than 4,000 feet, it has ranged since 1919 between 810 and 840. Below the zone of oxidation there appears to be no increase of fineness with depth. At the Kennedy mine, for example, the fineness of the bullion ranged from 815 to 834 in 1892, when the ore was won from the 1,600-foot level; in 1924, when the ore was obtained from the 4,000-foot level, it was 830.

On the average 70 per cent of the gold won at any mine in the Amador section of the Mother Lode belt is obtained by amalgamation, the other 30 per cent from the concentrates. The concentrates from the several mines range in value from $40 to $110 a ton. The amount of gold in the concentrates is in part determined by the particular milling practice employed—for example, whether the product is closely concentrated so as to eliminate most of the quartz or whether a low-grade product containing 80 per cent of sulphides is made. Aside from differences due to milling practice no systematic relations are discernible: concentrate from gray ore may be as high in gold as concentrate from quartz ore. Where the ore occurs within a vein the wall-rock sulphides are of low grade; but if the ore consists of mineralized country rock, the sulphides in the wall rock may be of as good grade as those in quartz ore.

WATER CONDITIONS UNDERGROUND

The quantity of water yielded by the mines decreases markedly with depth. Some mines have become completely dry; in others water in small quantities persists to the deepest levels attained. In 1924 water was entering the bottom level (then the 4,200-foot level) of the Kennedy mine from the footwall gouge. It was an unusually dry year, and in July only 72,000 gallons was pumped daily from this mine. At the adjoining Argonaut mine 150,000 gallons was lifted daily. The water pumped at the Plymouth mine in the summer of 1924 amounted to 30,000 gallons a day; in winter the quantity usually increases to 40,000 gallons a day. Of the quantity pumped only 600 gallons a day came from the bottom workings—those between the 3,400-foot and 4,000-foot levels.

The experience in deep mining on the Mother Lode coincides with experience elsewhere—namely, that mines become drier in depth. However, the control of fissuring rather than depth in determining the inflow of water is well shown at the mine of the Calaveras Copper Co. at Copperopolis, which is 14 miles west of the Mother Lode belt, in an exceedingly dry region. A crosscut driven on the 800-foot level into the 70-foot dike of granodiorite that occurs in the footwall of the copper lode encountered most unexpectedly a copious flow of water—60 gallons a minute.

QUARTZ VEINS

GENERAL FEATURES

The quartz veins are tabular masses of quartz dipping steeply eastward and striking between north and northwest. In any small exposure they appear to conform to the strike and dip of the inclosing rocks, but without known exception they cut these rocks in both strike and dip. The average divergence between the veins and the inclosing rocks is between 15° and 20°. It is most marked in those few parts of the gold belt in which the strata dip westward, as at Plymouth. As a result of this divergence the veins when followed longitudinally or in depth are found to cut diagonally through belts of rocks of differing character. Veins that crop out in greenstone may pass into slates in depth; "contact veins," so called because they are inclosed between a footwall of slate and a hanging wall of greenstone at the surface, pass at depth wholly into the greenstone; veins that crop out in slate may at depth become "contact" veins and at greater depth pass wholly into greenstone. Examples of all these conditions will be found in the descriptions of mines.

Where a vein passes either in strike or in dip from slate into greenstone or the reverse it is much deflected, or to use a term from optics, it is strongly refracted. This refraction is one of the most if not the most notable structural features of Mother Lode veins. The refraction is greatest in passing from the black slate to belts of massive greenstone—that is, belts that are made up of nonschistose breccia and lava sheets. The refraction is conspicuously shown at the Original Amador and Keystone mines. (See figs. 11 and 12.) The contact vein of the Keystone mine dips at the surface 75° E., but after entering the massive greenstone below the 800-foot level it flattens greatly, dipping as low as 20° over large stretches. The refraction is accordingly as much as 55°.

On entering schistose tuffs, rocks that are intermediate in strength between massive greenstone and slate, the veins are, of course, refracted much less. The rules governing the refraction obviously conform to those for the refraction of light; on entering the greenstone the vein is deflected toward the normal; on entering the slate it is bent away from the normal. The relations are so regular that they suggest the possibility of computing the index of refraction according to the formula $n = \dfrac{\sin i}{\sin r}$, in which n is the index of refraction referred to slate as unity, and i and r are the angles between the incident and refracted courses of the vein, respectively, and the normal to the contact. From the data obtained at the Original Amador, Keystone, and Central Eureka mines, the index at each of these mines thus computed is roughly 1.4. The extreme figure obtained (from the Argonaut mine) is 1.6. But whether 1.4 or 1.6 is used, the calculated dip of a vein in greenstone from its known dip in slate will differ only about 5°. The factors that preclude obtaining any great accuracy for the index of refraction are the considerable movement along the fissures, which has blurred the refraction at the contact between slate and greenstone, and the variant constitution of the greenstone belts and hence their diverse refractive powers.

The fissures along which the veins run are the results of reverse faulting. The maximum displacement as measured in the plane of the dip is 375 feet. The displacement first measured is that at the Argonaut mine, by Ransome,[52] who determined that the hanging wall had been thrust upward over the footwall for a distance of at least 120 feet. At that time the Argonaut was the only place along the Mother Lode where the displacement could be seen and measured. During the present investigation the displacement was measured also at the Original Amador, Gover, and other mines. At the Original Amador mine the hanging wall has been shoved 180 feet up over the footwall (see fig. 11), at the Treasure at least 300 feet, and at the Gover 375 feet.

A horizontal component in the faulting was measurable only in the Central Eureka mine, where it has caused a displacement of the walls of the vein amounting to 120 feet.

Adjacent to the veins the country rock, if slate or schist, has been bent or flattened against the veins. Hence in any local exposure, such as is seen at the face of a drift, the veins invariably seem to conform to the strike and dip of the inclosing rocks.

The slates near the veins are of highly characteristic appearance. They are buckled and kinked and corrugated, and their foliation surfaces have taken on an extraordinarily lustrous black polish. The greenstones adjacent to the veins, especially those veins that contain thick lenses of quartz, were highly shattered and are traversed by a network of quartz veinlets.

VEIN FILLING

The vein filling consists chiefly of a coarse milk-white quartz, which forms abruptly ending lenticular masses. At the ends of such lenses the ore body frays out into a stringer lode, consisting of slate penetrated by a large number of more or less parallel quartz stringers. The main body of solid quartz is generally accompanied on both hanging wall and footwall by a wide zone of quartz stringers, or stringer halo, as it may be called, as the closeness with which the stringers are spaced diminishes with distance from the main body of quartz.

The quartz filling is not crustified. It is commonly ribboned or banded, however, owing to the intercalation of slabs and filaments of black slate or of schist. These slabs may range in thickness from a few inches down to the thinness of paper. The paper-thin laminae may be most delicately and marvelously crenulated. The layers of quartz may be several feet thick or very thin, as shown in Plate 4, A.

The ribboning is invariably parallel to the walls of the veins, and therefore, like the veins, it makes an acute angle with the strike and dip of the inclosing rocks. The angular discordance between ribboning and bedding is of course most pronounced where the strata dip west, as at Plymouth, where the angle between ribboning and bedding is as much as 45°. The ribboning is probably due as a rule to successive openings and fillings of the vein fissures; in part it may be the result of the laminae of schistose slate that incompletely filled a vein fissure or adjoined the fissure having been pried apart by the force of crystallization of the growing quartz. In places the slate ribbons have been sheared by movement during a subsequent reopening of the vein.

The largest masses composed wholly of quartz occur in the greenstones. At the Keystone mine there are masses 54 feet thick. As a rule they carry not more than a dollar to the ton. Such great bodies of quartz are generally quite homogeneous, being without ribbon structure or crustification.

Inclusions of slate and greenstone in the quartz are common. They are all sharply angular, even those so minute as to be detectable only in thin sections under the microscope. Although all, without exception, have been strongly altered by the ore-forming processes, none have been replaced by quartz. They contain more pyrite and arsenopyrite than the quartz that incloses them. Not uncommonly the inclusions, those of slate as well as those of greenstone, are surrounded by a narrow fringe of ankerite. In Plate 4, B, is illustrated an assemblage of greenstone and slate inclusions; the narrow fringe of ankerite surrounding the large inclusion is well shown. The inclusion is nearly split in two by a veinlet (largely ankerite); possibly it was disrupted by the force of crystallization of the minerals as they grew in the veinlet.

Inclusions may locally be so abundant as to constitute an angular rubble cemented by quartz. Generally the inclusions are most abundant near the walls of the veins, but in places some are found far away from the walls and from other inclusions. Such, for example, were the isolated blocks of greenstone 1½ feet in diameter that occurred 8 feet above the footwall of the Original Amador vein, embedded in a pillar of coarse quartz ore 20 feet thick.

The quartz filling of many veins is clearly composite, having been brought in at successive times. This mode of origin is suggested by the ribbon structure of the

[52] Ransome, F. L., U. S. Geol. Survey Geol. Atlas, Mother Lode District folio (No. 63), p. 8, 1900.

veins, but more positive evidence is at hand. For example, the Calaveras vein, on Carson Hill, regarded as the local representative of the Mother Lode, plainly shows in places its composite origin, a dark quartz being cut by a network of veinlets of white quartz. The composite nature of the quartz filling of the ore shoot on the 4,200-foot level of the Argonaut mine is well shown in Plate 5, *A*. Many of the prominent outcrops of quartz, which by their size and persistence gave rise to the Mother Lode tradition, are of complex origin. Such a one is the Bull vein of the Eagle Shawmut mine, whose outcrop is an intricate network of quartz veins, strikingly emphasized by the way in which weathering has etched out the intervening material of silicified, ankeritized chlorite schist. (See pl. 5, *B*.)

GOUGE

The quartz veins, especially those in the slates, are generally accompanied by a black gouge, consist-

FIGURE 4.—Structural features of a quartz vein in slate. 1, Slate; 2, graywacke bed; 3, quartz; 4, gouge; 5, stringer zone in disturbed slate.

ing of triturated slate. In places the gouge is spotted with fragments of quartz, most of which have been reduced to a sugary condition. Gouge may occur on the hanging wall or the footwall or on both. In places it passes diagonally through the veins, and the suggestion arises that the abnormally thick portions of some of the veins may be due to the telescoping of one part of a vein on another by postmineral movement along such gouge zones.

The gouge pinches and swells and may attain a thickness of 8 feet or more. Thicknesses of 2 or 3 feet are most common. A notable habit of the gouge is its tendency to hug or wrap around the bodies of massive quartz; therefore a crosscut through the gouge adjacent to solid quartz bodies will commonly show

a wide zone of stringers in the adjoining slates. (See fig. 4.)

Where a vein is bordered on both the hanging wall and the footwall by gouge, the quartz of the entire vein is likely to have been crushed to a loose, sugary condition. It is the slates adjacent to the gouge that are particularly severely buckled, crumpled, and slickensided and have taken on a brilliant black polish.

In contrast to the general occurrence of gouge associated with the veins in slate, the great quartz vein of Carson Hill, which is inclosed in green schists and is well exposed in mine workings for 4,000 feet on the dip, is not accompanied by any gouge whatever.

The practical significance of the gouge is that it causes swelling ground, which makes heavy timbering necessary. In a few weeks 18-inch timbers will be reduced to matchwood, and unless drifts are continuously kept in condition and retimbered they soon cave in and are lost beyond recovery. The "swelling" is in part the result of the actual flowage of the gouge into the mine workings—an extrusion in the direction of least resistance—and in part is probably due to the imbibing of water by colloidal material in the gouge. At the Kennedy mine experience has shown that only the gouge causes swelling ground and that it is troublesome only where water has access to it.

ORE SHOOTS

GENERAL FEATURES

The gold is localized in the veins, and the ore shoot is generally short, being a mere fraction of the length of the vein in which it occurs. For example, the Empire shoot of the Plymouth mine, a shoot that has yielded $6,500,000, comprised only 450 feet of the 7,000 feet total length of the vein. A shoot that is 1,000 feet long is exceptional; 200 or 300 feet is more nearly an average stope length of Mother Lode ore shoots.

In spite of this small longitudinal extent some of the ore shoots have persisted from near the surface to 5,000 feet on the dip and at that depth are still profitable.

The pitch or rake of the shoots, as it is usually termed along the Mother Lode belt, is generally steep. The Empire shoot, however, pitched as low as 45° S. The Argonaut shoot taken in its entirety has a nearly vertical pitch, but a portion of it, from the 2,530-foot level to the 3,300-foot level, pitched 30° S. No one constant direction of pitch is common to all ore shoots along the Mother Lode belt: some shoots pitch north and others pitch south.

Many of the ore shoots did not crop out at the surface. The Central Eureka affords the most striking examples of this fact. Not until a depth of 1,100 feet on the vein had been reached was an ore shoot cut; in the deep levels of the mine an ore shoot on a different vein is now being worked, and the top of this shoot was on the 3,300-foot level.

Even in the shoots that have persisted most nearly continuously downward the value of the ore was not constant. Lean zones, some so lean as to be called barren zones, were met, and two or three such zones have been passed through in attaining a depth of 4,000 feet. Some were as much as 300 to 400 feet in vertical extent. Unfortunately the lean intervals have been fatal to many mines. From the way the term "barren zone" is used along the Mother Lode it is not always clear whether the barren interval is the result of the pinching of the vein to a gouge-filled fissure or of the quartz filling becoming unprofitable; barrenness of both kinds occurs.

The depth at which such barren zones or blanks in the vein may intervene is not at constant vertical distances below the outcrop, nor has it any relation to such an arbitrary datum as sea level. It is an individual matter with each mine. In fact, the puzzling problem of the ore shoots is peculiar to each mine. According to the manager of the South Eureka mine that mine was kept poor for 18 years by basing its program of exploration on that of its successful neighbor, the Central Eureka. After the fact became apparent that its ore shoots were in its footwall vein instead of in its hanging-wall vein, the South Eureka became prosperous.

The ore shoots are generally abrupt bulges in the veins and are considerably wider than the adjoining lean or barren portions. This feature is strikingly exemplified in the Plymouth, Central Eureka, Kennedy, and Argonaut mines. The filling of the veins in the shoots consists largely or wholly of quartz, which is either massive or ribboned. The ribboning, as already explained, is due to the intercalation of slabs and filaments of slate parallel to the walls of the vein between layers of quartz. Some writers distinguish between banded structure and true ribbon structure, but no such technical distinction is maintained here. As the ribbons of slate become more numerous, thicker, and more persistent the solid quartz vein grades into a stringer lode. Although ribbon quartz is regarded favorably as an indication of ore, many veins are ribboned beyond the limits of the shoots. This fact was noted at the Plymouth mine, where the superintendent, on contemplating a face of low-grade "ribbon rock," remarked, "It looks as if it ought to be ore, but it isn't." At the Central Eureka I have seen a face of finely ribboned quartz 40 feet wide, but it carried no gold, and even sectional assays showed no ore; and at the Princeton mine, near the south end of the Mother Lode belt, the vein is ribboned for hundreds of feet beyond the limits of the shoot. It must therefore be concluded that ribboning and gold content are independent features and that where they occur together it is merely by coincidence.

Many of the quartz masses that constitute the ore shoots fray out into stringer lodes and then become unproductive. This habit was seen in the Plymouth,

Central Eureka, Kennedy, and Argonaut mines and is a fact to be most carefully considered in any attempt to account for the origin of the gold shoots.

The only visible feature that distinguishes some but not all ore shoots from the remainder of the vein is a somewhat greater content of sulphides, commonly pyrite. As the sulphide content of the ore ranges only from 1 to 2 per cent, however, it is generally not a striking feature.

From the fact that the deep-level ore shoots—those in the Central Eureka, Kennedy, and Argonaut mines below a depth of 4,000 feet—carry very little sulphides, although the ore is of good and even high grade, it is my impression that the sulphide content decreases at depth, although there is no statistical evidence to prove that supposition.

Although the vein filling in the ore shoots consists largely or wholly of quartz, it is not true that all large masses of quartz in a given ore-bearing vein constitute ore. A vein containing an ore shoot may break up into a stringer lode, then pinch down to a gouge-filled fissure, and then expand abruptly to an impressive body of quartz, but this quartz may be barren. Some of the gravest disappointments in exploring Mother Lode veins have been due to this idiosyncrasy.

DISTRIBUTION OF THE GOLD WITHIN THE SHOOTS.

The gold in some of the shoots is distributed uniformly across the width of the vein, but in others, as proved by sectional assays, it is distributed unequally. In some the footwall portion is the most valuable, and in others the hanging-wall portion. At the Oneida, for example, the quartz within 6 or 8 feet of the hanging wall yielded from $30 to $40 a ton, but the whole vein, which was 10 to 40 feet wide, averaged $17.50 a ton.[63]

Detailed data showing the distribution of the gold in parts of the Empire vein are available; they constitute the only information of this kind from the Mother Lode known to me. Some of these data are reproduced in Plate 6, which not only gives the precise distribution of the gold but also shows the structural features of the vein. It appears from the diagram, as well as in others not reproduced here, that the gold in the Empire vein as a rule is distributed nonuniformly across the thickness of the vein, but although it tends to be more abundant along the footwall, yet in places the middle of the vein or the portion against the hanging wall contains as much or more gold than that near the footwall. If, as must be admitted, the gold is not systematically distributed with reference to the vein walls, nevertheless other systematic relations appear unmistakably. Unless there is a substantial body of quartz there is no ore; and where the vein becomes a stringer lode it becomes barren ($0.20–$1.60 a ton). Only in places adjacent to a

[63] Browne, J. R., Mineral resources of the States and Territories west of the Rocky Mountains for 1867, p. 74, 1868.

quartz shoot does a stringer zone constitute ore. A few assays of the gouge, listed in Plate 6, show a gold tenor ranging from $2.80 to $5.80 a ton, but in other parts of the mine the gouge carries as little as $0.40; where it grades into sugar quartz it may carry $7 a ton or more, its tenor evidently depending on how much quartz it has robbed from the ore shoots.

That in a ribboned quartz vein some one or more of the ribbons are not uncommonly far more productive than all the others was long ago noticed by Phillips.[54] But a systematic study of this relation appears never to have been made. In places it can be seen that some one ribbon carries most of the sulphides in the vein, and as sulphides generally indicate good ore it is likely that such ribbons carry most of the gold.

Sulphides are commonly contained in the slate filaments between the quartz bands or in the inclusions of slate or greenstone in the quartz. The visible free

FIGURE 5.—Thickness and assay value of ore shoot on the 2,150-foot level, Plymouth mine

gold may occur either in the slate filaments, or inter-grown with the sulphides, or inclosed in quartz. It was in the main deposited contemporaneously with the quartz and sulphides with which it is associated. Some of it has been sheared, either by movement along the fissure during one of the stages in the growth of the vein, or by postmineral movement after the vein had been completely formed.

The horizontal distribution of the gold in the Empire shoot on the 2,150-foot level is shown in Figure 5. The shoot on that level was 145 feet long, and its gold content ranged from $5 to $37 a ton. Its thickness averaged 9.5 feet, whereas the thickness in the next 175 feet along the vein north of the shoot averaged 5.4 feet and the gold content had sunk to $1 or $2 a ton. In Figure 6 the remarkable concentration of the gold

in the ore shoot is shown by means of a graph obtained by plotting the product of the assay value of the ore and the width of the vein in inches.

The south shoot of the Plymouth mine, so called because it was south of the Empire shoot, was 65 feet long on the 1,500-foot level, carried from $4.50 to $18 a ton, and averaged 8 feet in thickness. The adjoining unprofitable extension of the vein averaged but 4.4 feet in thickness for 100 feet north and 5.3 feet for 100 feet south of the ore shoot. On the level above, the 1,400-foot, the south shoot was considerably longer, attaining a length of 250 feet, and its average thickness here also was considerably more than that of the barren portion of the vein north of it.

The rule at the Plymouth mine, as shown by adequate quantitative evidence, is, therefore, that the ore shoots are considerably wider than the rest of the vein and contain more gold to the ton. Furthermore, the shoots consist more solidly of quartz, for as soon as the massive quartz bodies of which they are constituted fray out into stringer bodies they become poor in gold. Large bodies of quartz, however, do not necessarily make ore. These features of the ore shoots shown at the Plymouth mine are exemplified also in the Central Eureka, Kennedy, and Argonaut mines, and they are therefore regarded as sufficiently common to serve as a basis for some generalizations concerning the habits of Mother Lode ore shoots.

Many of the ore shoots are situated at the junctions of veins. Probably the most striking example was the old Empire shoot, although the information about this shoot, like that about most Mother Lode shoots, is none too full. This shoot was in the Empire vein at its junction with the Pacific vein; it pitched south at an angle of 45° down to the 1,500-foot level as measured in the plane of the Empire vein, having thus a pitch length of 2,000 feet. Now, the Empire vein strikes N. 10° E. and dips 70° E., whereas the Pacific strikes N. 5° W. and dips 85° E., a divergence of 15° in both strike and dip. This divergence persists in depth, although both veins become somewhat less steep. The inclination of the line of junction of the two veins, computed from the given data, is roughly 45° S., an agreement with the pitch of the ore shoot so remarkable that it can not be a mere coincidence but must be of genetic significance. Below the 1,500-foot level the Empire shoot lost the regularity that it had

54 Phillips, J. A., Notes on the chemical geology of the gold fields of California: Philos. Mag., 4th ser., vol. 36, p. 324, 1868.

above that level (see fig. 8), and this behavior appears in part explainable by the fact that the Pacific vein below the 1,500-foot level fails to join the Empire.

At many other mines ore shoots occur at junctions of veins, that is, where veins split into branches. Of course it is not always possible to ascertain whether we are dealing with an intersection or with a junction. The Kennedy vein (see p. 64) affords a notable example of an ore shoot at the branching of a vein, ore occurring both in the main vein and continuing in the two branches into which it splits.

The favorable influence of intersections and junctions is well known along the Mother Lode belt. Although intersections are likely places to explore for new shoots, it is true, on the other hand, that at some junctions large bodies of barren quartz have been found.

Shoots of gray ore occur in the wedge ends of green-stone masses that lie between converging fissure veins.

Shoots occur in the elbows of abrupt bends in the courses of veins. As such abrupt bends commonly occur at the passage from slate to greenstone and graywacke or the reverse, these shoots appear to be due to structural control by the wall rocks. Many ore shoots, however, are merely short lenticular expansions of the quartz veins in which they occur, and to all appearances their environment, structural and petrographic, differs in no way from that of the unproductive portions of the veins.

ORIGIN OF THE ORE SHOOTS

Previous ideas.—Early in the history of mining in the Mother Lode belt it became apparent that only parts of the veins carry enough gold to be profitably workable—that the ore occurs in "pay chimneys" or shoots. The first writers on the geology of the lode considered this limitation of the gold to definite ore shoots an intrinsic and universal feature of gold veins. Consequently, they regarded the occurrence of relatively barren spaces in veins as inevitable and the discovery of a rational explanation of the ore shoots as most unlikely. Raymond[55] puts the matter thus trenchantly:

Nature has not regulated her operations by commercial rules. To her the distinction between valuable and useless minerals, rich and poor ore, is unknown. * * * Of those fis-'sures which are filled with metalliferous material, some are rich enough to pay for working, and some are not. The same vein is in one spot rich and in another poor. These variations it is impossible to reduce to any general law, especially in the case

of gold and silver. The former of these is always and the latter frequently an accidental constituent in the vein material. One-thousandth of 1 per cent of gold would correspond with $5 gold per ton of ore—scarcely a paying quantity; two-thousandths of 1 per cent would be $10 per ton—enough under favorable circumstances to pay good profits. * * * It is manifest that changes so small as these can not have the importance naturally which we assign to them commercially, and that we are not likely, for instance, to find out by what law the proportion of gold varies from one to two thousandths per cent, crossing the arbitrary line of profit, which we, and not nature, have drawn.

Nevertheless, the problem of the ore shoots presses continuously and insistently on the attention of those guiding the exploration for ore in Mother Lode mines. The earliest generalization that has stood the test of time is that points of junction and intersections of veins are favorable to the occurrence of ore. This rule was announced by Hittell[56] as early as 1863, but remarkably enough later writers have not pointed out its importance. O. H. Hershey, however, has in private

FIGURE 6.—Concentration of gold in the ore shoot on the 2,150-foot level, Plymouth mine

reports to the Plymouth Consolidated Gold Mining Co. insisted on the value of this rule.

Fairbanks,[57] who in 1890 made the first comprehensive albeit hurried study of the Mother Lode belt, did not specifically discuss the origin of the ore shoots. No relation between the character of the walls and the poverty or richness of the quartz could be determined. The conditions influencing the occurrence of the gold "are certainly complex," he said, "differing greatly in different locations; the same character of ore is rich in one spot and poor in another, without any apparent reason for it."

Ransome,[58] in the Mother Lode folio, also does not specifically discuss the origin of the ore shoots. He

[55] Raymond, R. W., Statistics of mines and mining in the States and Territories west of the Rocky Mountains for 1869, p. 456, 1870.

[56] Hittell, J. S., The resources of California, p. 274, 1863.

[57] Fairbanks, H. W., Geology of the Mother Lode region: California State Min. Bur. Tenth Ann. Rept., pp. 23–90, 1890. A condensed account is given in Am. Geologist, vol. 7, pp. 209–222, 1891.

[58] Ransome, F. L., U. S. Geol. Survey Geol. Atlas, Mother Lode District folio (No. 63), p. 8, 1900.

found, in harmony with Fairbanks's conclusion, that with the possible exception of serpentine there is no indication that the petrographic character of the wall rock "has any direct influence upon the richness or poverty of the vein." The most favorable rocks are those having cleavage or schistosity. Stringer leads and impregnated zones of fissile and schistose rocks are more uniformly productive than the large veins of solid quartz. This last rule requires some modification, for in Amador County, where the veins of solid quartz break up or fray out, as it were, into stringer lodes they become too poor to work. Ransome also found that the ribbon ore "is almost invariably of more than average value," but this generalization is far from true.

In his bulletin on the Mother Lode region of California Storms [59] combats the long-ingrained belief that the Mariposa slates are essential to the occurrence of good ore. He says:

Without exception, the large number of accessible mines prove that the Mariposa clay slates form no really important feature as related to ore deposition, while thus far in Amador County, south of Plymouth, all development confined to the typical clay slates of the Mariposa beds has proven the fissures in that formation valueless, and these developments reach many thousands of feet of shafts and drifts.

The most productive ore bodies of Amador County, according to Storms, are invariably associated with a peculiar slaty rock derived from the metamorphism of an augitic tuff. Even though this generalization has been carefully limited to Amador County, it is still too sweeping, for the bonanza ore bodies of the Plymouth and Keystone and of the deep levels of the Central Eureka have been wholly incased in Mariposa black slates.

In 1904 W. A. Prichard,[60] who had been superintendent of the Keystone mine, emphatically asserted the favorable influence of the Mariposa black slates on the amount of gold in the veins. Lodes that cut across interbedded slate and greenstone are said to vary remarkably in tenor, in correspondence with the amount of carbon in the slates. In addition to the factor of carbon content, structure determines the richness of the veins. "The most valuable veins of the Mother Lode," Prichard stated, "are those occupying fissures whose course is zigzag or winding."

The belief that ore shoots occur in definite zones in depth, which is more or less current along the Mother Lode belt, apparently was first publicly expressed in 1894. According to Emmens,[61]

It is now recognized by every mining or geological expert who has examined the mines of Amador County that there are two zones of gold ore in the Mother Lode, one extending from the surface to a depth of usually 600 or 700 feet, and the other commencing at about the 1,100-foot level and extending to a depth at present unlimited.

As a generalization this is far from true, for the definite zonal arrangement does not exist, but the element of fact in it is that the ore shoots are discontinuous, with intervening lean or barren zones, in depth as well as longitudinally.

In 1911 Storms [62] recurred to the problem of the ore shoots. He sketched briefly the history of the ore shoots of the chief mines along the belt. From that history he concluded that all the profitable mines of the Mother Lode had zones of good ore and zones that were of too low grade to be profitable, and that these zones were not uniformly distributed. Further, he concluded that

When all the available data are collected and careful comparison made, it may be found that there is more regularity in the occurrence of ore shoots on the Mother Lode than a casual glance may indicate. To me it seems that proper geological conditions are of greater importance than all other considerations, and that where these are present there is every reason to anticipate the recurrence of rich ore in the fissures to a depth as great as lies within human skill and mechanical skill to reach.

But what the "proper geological conditions" are is not disclosed.

Maclaren [63] has strongly maintained that the carbonaceous Mariposa slates are essential to the persistence of ore in depth. This factor is thought to account for the greater depth to which ore extends in Amador County than it apparently does at Angels Camp, where the veins are inclosed in amphibolite schist. But no proof is supplied for this proposition. According to Maclaren,

Many mines have certainly "bottomed" the ore in given fissures at depths less than 2,000 feet, but it often happens that two or more parallel lodes occur within the Mariposa slates and that when one becomes barren, a hanging-wall or footwall lode may carry ore to much greater depths. In few auriferous regions is crosscutting from wall to wall of the lode channel more necessary; in few has less been done than along the Mother Lode.

In conclusion, it appears that the views on the ore shoots are highly conflicting. Generally the specific examples on which the rules have been formulated are not given, so that it is impossible to review the evidence and to weigh the validity of the conclusions reached. Is a given generalization based on one or on more than one occurrence? Is it the only possible interpretation of the evidence? Take the oft-asserted efficacy of the carbon in the black slates in producing high-grade ore: Has the carbon (carbonaceous matter, methane, or similar constituent) actually caused the

[59] Storms, W. H., California State Min. Bur. Bull. 18, pp. 15–19, 1900.

[60] Prichard, W. A., Observations on Mother Lode gold deposits, California: Am. Inst. Min. Eng. Trans., vol. 34, pp. 454–466, 1904.

[61] Emmens, S. H., The Mayflower mine, California: Eng. and Min. Jour., vol. 57, pp. 173–174, 1894.

[62] Storms, W. H., Possibilities of the Mother Lode in depth: Min. and Sci. Press, vol. 103, pp. 646–648, 1911.

[63] Maclaren, Malcolm, The persistence of ore in depth: Cong. géol. internat., 12e sess., Compt. rend., pp. 300–301, 1914.

more abundant precipitation of gold, or was the more abundant deposition due to the solutions entering an abruptly wider part of the fissure, which slowed down their progress and thus facilitated the deposition of the gold?

The strongest impression that the review of the earlier literature leaves is the almost complete lack of precise quantitative data on the distribution of the gold, such as is given by assay charts, and the absence of attempts to correlate the localization of the gold by means of detailed data with the structural features of the veins and with the rocks that inclose them.

Influence of wall rocks.—That the black Mariposa slates have exerted a favorable influence on deposition is supported though not proved by some facts. Of the $17,000,000 produced by the famous Keystone mine, the bulk was yielded by the veins that were inclosed in slate, whereas the upper part of the "contact" vein worked in recent years, which has a hanging wall of greenstone, is of low grade, and the part that lies wholly in the greenstone is poorer still. In the Central Eureka mine the vein that contained the bonanza-ore shoot, extending from the 1,100-foot to the 1,900-foot level, became barren after it had entered the greenstone below the 3,000-foot level. The new high-grade shoot in that mine, whose top is at the 3,300-foot level, is inclosed in slate, whereas the adjacent contact vein, having a hanging wall of greenstone and a footwall of slate, is barren.

In the South Eureka, a vein in passing from greenstone into slate abruptly increased in gold content from $4.60 to $10 a ton. The upper portion of the Argonaut-Kennedy vein, extending from the outcrop to a depth of several hundred feet, was of too low grade to work, but the vein improved in depth on entering the slate. As this vein cuts through alternating belts of slate and greenstone, it is possible that there is a partial correlation between the alternation of rich and poor zones encountered in depth and the alternation of slate and greenstone. But precise data to test this supposition are not available to me.

The foregoing summary appears to comprise all the positive evidence that can be adduced from the veins themselves that the slates have exerted a favorable influence. Probably the safest conclusion that can be drawn from the evidence is that the greenstone where it forms both walls of a given vein has exerted a somewhat unfavorable influence on the gold content. On the other hand, as if partly to offset this unfavorable verdict against the greenstone, is the fact that in places the greenstone has been transformed by the action of the gold-carrying solutions into large valuable bodies of the so-called gray ore.

Even if it is granted that the slates have exerted a beneficial influence on the tenor of the veins in them,

it is clear that they are not essential to the occurrence of ore bodies, for extremely productive veins have been worked in amphibolite schist at Angels Camp, in amphibolite and chlorite schists at Carson Hill, and even in ankeritized serpentine at the Rawhide mine. Evidently the composition of the wall rock was not the controlling factor in determining the occurrence of ore in the Mother Lode belt.

The inclusions of wall rock in the quartz veins, whether they are slate, graywacke, or greenstone, commonly carry far more sulphides than the inclosing quartz. Thus it is plain that the country rock has exerted a precipitating effect on the ore-forming solutions. Therefore it would seem logical to expect that where the ore-forming solutions came most intimately into contact with the wall rocks, as in the narrower parts of the veins and especially in the stringer lodes, they would be richest in sulphides and in gold. But this expectation is the reverse of the real state of affairs: the ore shoot is wider than the remainder of the vein, it consists predominantly of quartz, and where it breaks up into a stringer lode it becomes barren. Hence again the conclusion is enforced that the chemical nature of the wall rock can not have been the factor determining the gold tenor in the veins of the Mother Lode belt.

Proposed hypothesis.—Intersections or junctions of veins are clearly one of the factors that favored the forming of ore shoots. Such intersections occur where veins fill auxiliary fractures in the zone of fracturing that accompanies the main fault. Some veins branch where they pass from rock of one kind to rock of another, but the cause of branching is as a rule not explicable and hence the place of branching can not be foretold in advance of exploration. Why vein junctions should have favorably influenced the formation of ore shoots in the quartz veins is not clear from any facts yet known concerning the Mother Lode. The origin of the gray-ore shoots, however, appears to be rationally explicable by the fact that the mass of greenstone occupying the wedge end between two fissures that intersect at an acute angle was thoroughly shattered, and the ingress of the gold-depositing solutions was thereby facilitated.

The ore shoots are most probably accounted for on the hypothesis that they occupy the main channels along which the ore-forming solutions flowed to the earth's surface. The ascending currents would tend to be localized in these channels for the following reasons: Thermal waters issue at the earth's surface not along the full length of a fissure but only at definite orifices. Phillips [64] pointed out that this is the fact at Steamboat Springs, Nev., and it is char-

64 Phillips, J. A., Notes on the chemical geology of the gold fields of California: Philos. Mag., 4th ser., vol. 36, pp. 426–427, 1868. See also Becker, G. F., Quicksilver deposits of the Pacific slope: U. S. Geol. Survey Mon. 13, pp. 338–339, 1888.

acteristic of fissures at Karlsbad, Bohemia, and at many other places. Such localized efflux of the ore-forming solutions would necessarily determine directions of maximum flow in these solutions as they rose toward the surface, and the channels followed by the maximum flows would determine the position of the ore shoots in veins of hydrothermal origin, as indicated apparently first by DeLaunay.[65] The same idea was advocated by Grout,[66] and recently it has been carefully developed by Quiring.[67]

The other factor determining the directions of active ascent of the ore-forming solutions would be the structural features of the fissure—the size and continuity of the open spaces. Owing to the fact that the total displacement along the fissure was the cumulative result of successive intermittent movements, it is probable that the size and continuity of the cavities varied from time to time.

Only in the continuous passageways connecting the point of entrance of the solutions into a fissure with the points of discharge at the surface would there be upward-moving currents; elsewhere in the fissure the solutions would be nearly or quite stagnant. Past any given point in the trunk channels there would be a continuous flow of gold-bearing solutions that were saturated with gold brought up from great depth; consequently there could be a continuous deposition of gold. In the stagnant solutions filling the parts of the fissure distant from the trunk channels the gold necessary to maintain a concentration sufficient to cause precipitation could be supplied only by diffusion, an extraordinarily slow process compared with transportation by a moving solution. Inasmuch as the silica for the quartz in the veins appears to have been supplied largely by the wall rocks, the stagnant solutions became more easily supersaturated with quartz than the solutions in the principal conduits. Consequently the parts of the fissure distant from the main channels became filled with quartz that carries less gold than that deposited in the main channels.

That the ore shoots are as a rule thicker than the unproductive parts of the vein has been pointed out. Although the abrupt bulging of an ore shoot is the cumulative result of the mode of growth by successive reopenings of the fissure, it is probable that at any stage of its growth the ore shoot was thicker than the adjacent parts of the vein. In the wider parts of the fissure the rate of precipitation of gold might conceivably be increased, because there the movement of the solution would be slackened. Perhaps this cause may be sufficient to account for the few additional thousandths of 1 per cent of gold that occur in the ore shoot. The absolute increase in the quantity of gold in the

ore shoot is indeed small, but the relative increase is large—five, ten, or even twenty fold. The essence of the problem of the ore shoot can be put in the query, What factor or factors increase the rate of precipitation of the gold relatively to that of the quartz? The answer is by no means clear. It is not obvious in what way the diminution in the speed of flow of the ore-depositing solution increases the precipitation of gold relatively to that of the quartz. The hypothesis here proposed is based on the premise that the quartz in the vein filling is supplied from the wall rocks as a result of the decomposition of the silicates by carbon dioxide, by "lateral secretion," and the gold is derived from a deep-seated, probably magmatic source. The quartz is therefore delivered to the growing vein essentially by diffusion through the stagnant body of solution permeating the wall rocks; hence the rate at which it is delivered is governed by the laws of diffusion—namely, that the quantity delivered is proportional to the difference in concentration and that it increases with the temperature. Both of these factors are independent of the width of the vein. On the other hand the gold, brought to the place of deposition by the ascending thermal solution, is responsive to the changes affecting that solution, and its rate of precipitation will consequently be influenced by changes in the width of the fissure. In the slower moving current there will be more time for the gold to settle out. Thus because of the unchanged rate of deposition of the quartz and the increased rate of deposition of the gold the wider portions of the veins will become enriched in gold.

Furthermore, there is a suggestion that the accelerated rate of gold deposition in the ore shoot may lead to an impoverishment or undersaturation of the rising solutions in gold where the vein channel contracts, and they must flow upward some distance until loss of temperature and reduction of hydrostatic head cause renewed deposition. Possibly the alternation of lean or barren zones, where not due to the pinching of the vein to a gouge fissure, is thus to be explained.

The hypothesis that the ore shoots were formed in the trunk channels of circulation, the gold having come from a deep-seated source and the gangue mainly from the adjacent wall rocks by lateral secretion, will also account satisfactorily for the fact that in some ore shoots the gold is very unequally distributed. As the veins were formed by the filling of cavities intermittently enlarged, it is quite probable that the lean sections were filled while the developing ore shoot was temporarily not in the main pathway of the ascending solutions. If the switching of the ore-forming solutions from one pathway to another were frequently repeated during the growth of the vein, the resulting distribution of the gold would be highly erratic and would defy rational explanation.

[65] DeLaunay, L., Traité de métallogénie, vol. 1, p. 130, 1913.
[66] Grout, F. F., The localization of values or occurrence of shoots in metalliferous deposits: Econ. Geology, vol. 11, pp. 395–397, 1916.
[67] Quiring, H., Thermenaufstieg und Gangeinschieben: Zeitschr. prakt. Geologie, Jahrg. 32, pp. 161–171, 1924.

ORE BODIES OF MINERALIZED COUNTRY ROCK

MINERALIZED GREENSTONE ("GRAY ORE")

Nature and occurrence.—Gray ore is the name given to ankeritized greenstone that carries sufficient gold to be worked at a profit. The term is not mentioned in the Mother Lode folio, as ore of this kind does not appear to have been worked when that folio was written; it is apparently first mentioned by Prichard.[68] In recent years gray ore has been the mainstay at the Fremont, Bunker Hill, Treasure, and Keystone mines, and to a less extent at the Original Amador mine. Similar material, called gray ore but carrying insufficient gold to be workable, occurs at the Argonaut. Immense quantities of gray ore have been extracted from the Royal mine, at Hodson, Calaveras County, many miles west of the Mother Lode belt.

The gray ore, as its name implies, has a gray color. It is composed largely of fine-grained ankerite, with more or less sericite, albite, and quartz and 3 to 4 per cent of pyrite and arsenopyrite. It is commonly traversed by veinlets of quartz containing ankerite and albite. In composition the gray ore is like the ore formerly worked at Angels Camp, which was formed from the alteration of amphibolite schist, but it is unlike that ore in appearance. The difference is due to the fact that the gray ore was formed by the hydrothermal alteration of pyroclastic breccias and tuffs of augitic greenstone. The fragmental structure of the breccias and tuffs has not been obliterated by the transformation to ore; in places it has even been strikingly emphasized. Even the microscopic details have been preserved, such as the zonal structure in the augite crystals, although these crystals have been completely altered to ankerite. The gray ore ranges from massive to schistose, depending on the structure that it has inherited from the greenstone.

The gray ore occurs in shoots that lie adjacent to quartz fissure veins, in the footwall or hanging wall of the veins, though not necessarily in contact with them. For those of the gray-ore bodies that have a quartz vein as one wall, the other boundary is generally defined only by assays that show where the gold tenor decreases below ore grade. Outward from the gray-ore body the rock becomes greener on account of the less thorough replacement of the greenstone by ankerite. Although in general the boundaries of the gray-ore bodies are "assay walls," yet where slaty beds occur interbedded in the greenstone they constitute sharply defined walls.

Many of the shoots are large; in addition to this desirable feature the ground stands well and requires little timbering. The largest gray-ore body at the Fremont mine was 300 feet long, 16 to 70 feet wide (averaging 40 feet), and several hundred feet on the dip. It formed the wedge end of a mass of greenstone that lay between two converging fissures. The large bodies of gray ore in the Bunker Hill mine occurred in a similar position.

The size of the gray-ore bodies is not proportional to the thickness of the quartz veins with which they are associated. The large shoot of gray ore in the Fremont mine adjoined a thin quartz vein, and the gray ore in the Keystone mine did not adjoin the quartz vein where the vein was 54 feet thick but where it was much thinner.

A notable feature of some of the gray-ore bodies, doubtless of genetic significance, is that although the quartz in the adjacent vein is barren of sulphides the gray ore nevertheless carries abundant pyrite and arsenopyrite.

The gold content of the gray-ore bodies is reported to be spotty but is said to range as high as $20 a ton. Large bodies averaging $8 or more have been mined. The value of the ore can not be estimated on inspection, and remarkably enough no one at the mines where it is being extracted claims that he can do so.

Origin.—The gray ore has resulted from the replacement of greenstone, especially of the varieties rich in augite. Slaty varieties, containing an admixture of black slate substance, were distinctly unfavorable to replacement. The gray ore from the several mines as seen in the many thin sections examined is essentially solid ankerite rock. The ankerite that has replaced the augite crystals or grains is of a deeper color than that which has replaced the matrix in which they were embedded; as a consequence the original texture of the greenstone as a rule is still plainly apparent.

Drastic chemical changes resulted from the replacement. Sulphur and arsenic were added, forming pyrite and arsenopyrite. Carbon dioxide was added in large quantity, and it completely displaced the silica in the augite and other silicates. As the greenstone originally contained 50 per cent of silica, it is manifest that enormous quantities of silica were carried away from the wall rocks. The amount is amply sufficient to account for all the quartz in the veins.

Rarely the replacing solutions carried off more material than they deposited and thereby produced a somewhat vuggy rock. Gray ore of this kind was locally formed in the Keystone mine, containing vugs half an inch in diameter that are lined with small flat crystals of ankerite.

The gold that has been added to the ankeritized greenstone has traveled a long distance from the quartz veins, which represent the filled conduits up which it was brought. This is contrary to its usual habit, which is to stay within the vein walls. In places, as previously mentioned, the conditions that prevailed within a vein did not allow the gold and the sulphides to be precipitated, but the conditions in the adjacent greenstone did permit their precipitation. These relations suggest that high temperature was the determining condition, and that this high temperature supplied the

68 Prichard, W. A., Observations on Mother Lode gold deposits: Am. Inst. Min. Eng. Trans., vol. 34, p. 465, 1904.

necessary driving force to cause the gold to diffuse through the solutions that permeated the wall rock of the vein.

MINERALIZED SCHISTS

Mineralized schists, originally amphibolite and chlorite schists, have been mined at certain localities, notably at Angels Camp and Carson Hill. They are generally of low grade, carrying from $2 to $3 a ton. Although formerly worked extensively, as at the Melones mine, on Carson Hill, they can not now be worked at a profit. Even before the severe decline in the purchasing power of gold most of them barely yielded a profit.

The immense body of mineralized schist in Eldorado County known as the Georgia Slide, long regarded as marking the north end of the Mother Lode, was deeply rotted and was worked on a large scale by hydraulicking.

The pyrite-bearing sericitic-schist ore of the Gold Cliff mine, at Angels Camp, can be taken as representing much of the ore formerly mined in that district. The ore bodies consist of highly altered schists through which ramify quartz and ankerite veinlets that run generally transverse to the foliation. The schists carry several per cent of pyrite disseminated in sharp cubes, which range from microscopic dimensions to half an inch on an edge. The mineralization of the schists extends at least 100 feet from the footwall of the fissure from which it spreads out. The stringers rarely exceed a few inches in thickness; the thicker ones consist chiefly of coarse white quartz with minor ankerite, and those less than half an inch thick are likely to consist wholly of ankerite. The whole face of mineralized schist exposed in the immense pit at the Gold Cliff mine is reported to average 80 cents a ton over a width of 100 feet.

The mineralized schists are of various kinds, owing in part to initial differences of compostion and in part to differences in the degree of alteration produced by the mineralizing solutions. Under the microscope the mineralized schists are seen to consist predominantly of ankerite, with subordinate pyrite, sericite, quartz, and albite and accessory rutile. Free gold inclosed in pyrite was detected in one thin section. In the transformation to ore the schist has been thoroughly altered by the newly formed ankerite and sericite. Its schistose structure has remained intact, however, and, although obviously it has been derived from a member of the amphibolite schist, proof can not be brought to show whether it was originally a chloritic or an amphibolitic schist.

Although the schist ore bodies in general are low in gold, there are nevertheless some marked exceptions. The most conspicuous exception was the hanging-wall ore body of the Melones mine, which yielded large quantities of high-grade ore. This ore body formed a lens of auriferous pyritic schist lying against the hanging wall of a thick quartz vein that carries only a dollar in gold to the ton. This schist is netted with veinlets of quartz or ankerite or both and is highly ankeritic and somewhat sericitic, though the sericite is more prominent to the eye than its real abundance warrants. The original mineral composition of the schist has been completely transformed by mineralizing solutions. It is clear that schist of more than one kind was thus transformed, for the ends of the ore lens are blunt and break across rocks of several different kinds. The deposition of the high-grade ore can not be ascribed to the favorable influence of a particular kind of schist.

In the footwall of the same great quartz vein are immense bodies of low-grade pyritic schist that is netted with quartz veinlets. This schist was ore and was mined for many years, although it averaged less than $2 a ton. In appearance this low-grade material differs in no way from the high-grade ore that occurs in the hanging wall of the quartz vein. Like the high-grade ore it is a pyritized, ankeritized, and sericitized schist. It is true that the hanging-wall ore, besides its much higher gold content, carries in places a notable amount of molybdenite, a mineral that does not occur in the footwall ore; but this molybdenite content can hardly be significant. At Carson Hill we have the remarkable fact that there are associated with one and the same barren quartz vein footwall ore bodies of low-grade mineralized schist and a hanging-wall ore body of high-grade mineralized schist that is one of the richest ore bodies ever mined on the Mother Lode belt.

Microscopic examination of high-grade ore, carrying $50.20 a ton, from the 4,125-foot level of the Melones mine showed it to contain 90 to 95 per cent of ankerite ($\omega = 1.70$), the remainder being pyrite, sericite, quartz, albite, and rutile. The pyritic schist mined from the footwall ore bodies is similar in composition, as its megascopic appearance suggests, except that it contains less ankerite and more sericite. Some of this ore was derived from the alteration of a clinochlore schist, and the original green color of the schist diminishes and disappears as the Bull vein is approached, owing to the increasing amounts of sericite and ankerite formed at the expense of the clinochlore.

MINERALOGY

In the following paragraphs are described the modes of occurrence of the more important minerals, chiefly those minerals that are associated with the ore bodies, and some of their more noteworthy features are given. For further details on the mineralogy of the district the valuable work on minerals of California by A. S. Eakle [69] may be consulted. The minerals are described in alphabetic order.

Aegirine-augite.—Aegirine-augite is a microscopic constituent of the albitite and albitite porphyry

---

[69] California State Min. Bur. Bull. 91, 328 pp., 1923.

described on pages 21–22. It is identified by its negative elongation, deep-green pleochroism, and rectangular cleavage. Its extinction $a \wedge c$ is 16°, and its maximum refractive index is approximately 1.76. Its composition, as roughly deducible from the chemical analysis of the rock in which it occurs, includes 25 per cent by weight of the diopside molecule, but its optical properties indicate an even higher content of the diopside molecule, probably nearly 40 per cent, as determinable from the composition diagram given by Winchell [70] for the alkali pyroxenes. Much of the aegirine-augite has been altered to radiating groups of acicular riebeckite, which do not exceed 2 millimeters in diameter.

Albite.—The feldspar albite is a common constituent in the ores and altered wall rocks. It is rare in the thicker quartz veins, but at many places it is abundant in the veinlets that traverse the adjacent wall rocks. Albite-bearing quartz veinlets may contain free gold, as at the Georgia Slide mine. Albite may be so abundant as to form veinlets consisting wholly of a drusy aggregate of imperfect crystals. Veinlets of this kind occur in the black slates at the Sherman mine, near Placerville. Microscopically the albite shows multiple twinning, and its optical properties, including symmetrical extinction of 18°, prove it to be a nearly pure sodium feldspar.

Practically any thin section of altered wall rock will show some albite. It occurs in hydrothermally altered rocks of the most diverse kinds—greenstone, amphibolite, slate, and even quartzite. Albitization has been so thorough in places as to produce fine-grained white rocks that might well be called albite jasperoids. Such extreme albitization has locally affected greenstone, amphibolite schist, and chlorite schist.

Albite is extraordinarily abundant in the augite melaphyre lavas, tuffs, and breccias and in the keratophyres. How much of this albite is pyrogenetic and how much of it is derived from more calcic plagioclase as a result either of pneumatolysis accompanying the eruption of those rocks or of later albitization has not been satisfactorily determined.

Allanite.—Allanite is a notable accessory constituent in a hydrothermally altered rock that occurs in the east crosscut of the 300-foot level of the Ford mine, in Calaveras County. It is in microscopic brown crystals which have been broken and moved by subsequent deformation. It has produced pleochroic haloes in the chlorite that it touches.

Allanite appears also as an accessory mineral in the biotite granite that occurs in the metadiorite mass east of San Andreas. It is not improbable that the allanite-bearing rock in the Ford mine is an offshoot from this mass or is a related dike.

Ankerite.—Under ankerite are included those dolomites containing more or less iron carbonate in solid solution as $CaFe(CO_3)_2$.[71] Such iron-bearing dolomites are often referred to as ferriferous dolomites or ferrodolomites, but the name ankerite is preferred here, for it was applied as early as 1867 by Silliman to the iron-bearing dolomite that is common in certain ores of the Mother Lode belt, and it is well known along the gold belt.

Ankerite occurs in enormous abundance. Most impressive are the great masses of ankerite rock that have resulted from the replacement of serpentine. Just west of Coulterville, for example, is a belt of ankerite rock 300 to 500 feet wide. Other notable occurrences are at Carson Hill, the Rawhide mine, Quartz Mountain south of Jamestown, the Eagle Shawmut mine, the Mary Harrison mine, and the Josephine mine. The ankerite in these occurrences is a coarse white carbonate. Replacement of the serpentine has commonly been so thorough that the ankerite rock resembles a foliated marble, more or less stained green by mariposite.

On oxidation the ankerite yields a highly limonitic gossan. As Silliman remarked long ago: "Before decomposition this triple carbonate of lime, magnesia, and iron is brilliantly white, and its real chemical character would never be suspected."

Less prominently, because of finer grain and because more admixed and intergrown with other minerals, ankerite is nevertheless extremely abundant in other hydrothermally altered wall rocks—amphibolite schist, augitic greenstones, and even slates, graywackes, and quartzites. The metasomatic ankerite in the slates, graywackes, and quartzites is in idiomorphic rhombs.

Ankerite occurs also to a minor extent in the quartz filling of the veins proper. For example, it incrusts the inclusions of black slate in the quartz ore on the 3,800-foot level of the Plymouth mine. It is commonly associated with albite, and like albite it is more abundant relatively to quartz in the veinlets that extend from a large vein into the adjacent country rock.

Under the microscope the ankeritic nature of the carbonate so abundantly associated with the Mother Lode veins is commonly shown by the twinning bands, which bisect the obtuse cleavage angle. This optical test distinguishes dolomite and ankerite from calcite and magnesite. Proof of the ankeritic nature of the carbonate is given here, for Lindgren [72] denied that ankerite occurs along the Mother Lode and thought that the carbonate was a mixture of varying composition, ranging from calcite to magnesite and commonly containing considerable iron, the magnesite, however, predominating. Chemical tests show abundant iron, and the amount of iron present in the ankerite can roughly be approximated by determining the index ω.[73] For pure dolomite (containing only 0.11 per

[70] Winchell, A. N., Studies in the pyroxene group: Am. Jour. Sci., 5th ser., vol. 6, p. 519, 1923.

[71] Doelter, C., Handbuch der Mineralchemie, Band 1, p. 371, 1912.
[72] Lindgren, Waldemar, Characteristic features of California gold-quartz veins: Geol. Soc. America Bull., vol. 6, pp. 234–235, 1895.
[73] Ford, W. E., Studies in the calcite group: Connecticut Acad. Arts and Sci. Trans., vol. 22, pp. 11–248, 1917.

cent $FeCO_3$) $\omega = 1.680$;[74] for the ankerites of the Mother Lode ω generally ranges from 1.705 to 1.76.

Ankerite whose index $\omega = 1.705$, from a chloritic ankerite rock, was carefully analyzed by W. W. Smith, of Yale University, with the following results:

Analysis of ankerite from the Eagle Shawmut mine

[W. W. Smith, analyst]

	Percent-age	Molecular proportion	
CaO	32. 13	0. 573	
MgO	16. 57	. 411	1. 044
FeO	3. 85	. 054	
MnO	. 43	. 006	
CO₂	46. 14	1. 048	
Insoluble	. 98		
	100. 10		

Computed in terms of the mineral molecules present, the analysis shows:

$CaMg(CO_3)_2$	75. 6
$CaFe(CO_3)_2$	11. 7
$CaMn(CO_3)_2$	1. 3
$CaCO_3$	10. 2
	98. 8

The analysis shows that the ankerite contains a large amount of $CaCO_3$ in excess of the 1 : 1 ratio for dolomite. As the analyzed material was homogeneous, this excess $CaCO_3$ is present as an isomorphous constituent, as pointed out by Foote and Bradley[75] in their investigation of dolomite and confirmed by Ford[76] from a study of more than 100 analyses of dolomites.

Antigorite.—Those serpentines whose crystallinity is apparent to the unaided eye are found microscopically to be made up chiefly of antigorite. This identification is based on the lamellar habit, fine cleavage, positive elongation, straight extinction, and refractive indices of the mineral. As determined on antigorite from the serpentine mass northwest of the Ford mine, $\gamma = 1.573$ and $\alpha = 1.562$, and the birefringence is approximately .010.

Arsenopyrite.—Arsenopyrite is fairly common in the ores of the Mother Lode mines. It does not occur in all of them, however. At some mines its appearance in the quartz coincides with extraordinarily high content of gold, but in most it is apparently without significance. It is the most abundant sulphide next to pyrite. It occurs embedded in the quartz of the veins, where it is either crystallized or without crystal boundaries, and invariably as sharp, well-formed, lozenge-shaped crystals in the country rock near the veins, whether slate, graywacke, or greenstone. The

crystals cut sharply across the foliation of the rocks and are clearly of metasomatic origin.

It occurs in the gray ore of the Fremont and Bunker Hill mines. In the Fremont mine, according to the superintendent, arsenopyrite indicates high gold content, but in the adjoining Gover mine it is worthless. It occurs in the unique body of highly sulphidic ore mined in the deeper levels of the Eagle Shawmut mine. It occurred sparsely in the ore of the Plymouth mine, as far down at least as the 2,300-foot level, but it is extremely rare or does not occur in the bottom levels. Whether this disappearance of arsenopyrite in the deeper levels is an effect of depth, such as has taken place in the deeper mines of Cornwall, or is merely fortuitous, the evidence is too scant to decide. It is probably fortuitous, however, for arsenopyrite, intergrown with coarse gold, occurs in the much deeper levels of the Central Eureka mine, 7 miles south of the Plymouth.

Arsenopyrite was particularly abundant in the ore worked in the Keystone mine in 1915. This ore was partly "gray ore" and partly a fault breccia of slate and greenstone containing much metasomatic arsenopyrite and pyrite. But this ore, in spite of its high content of arsenopyrite, carried only $2.69 in gold to the ton.

In the belt of remarkable pocket mines extending from a point a few miles south of Jackson to Mokelumne River coarse gold is intergrown with massive arsenopyrite. Masses of gold more or less crystalline, weighing as much as half an ounce, were seen by the writer.

A cobalt-bearing variety of arsenopyrite (danaite) has been found at the Josephine mine; in the oxidized zone it gave rise to the peach-blossom pink cobalt bloom, erythrite.[77]

Bastite.—The serpentine near some of the mines (Rawhide, Clio, and Mary Harrison) contains numerous platy masses of bastite. They are embedded in a dark aphanitic matrix, and consequently they cause the serpentine to have an apparent porphyritic texture. They have a perfect cleavage, which reflects light with a brilliant, almost metallic luster. They are serpentine pseudomorphs after an orthorhombic pyroxene. Examined microscopically, cleavage plates of the bastite were found to show central emergence of the negative bisectrix. The angle 2E of bastite from the Clio mine, Tuolumne County, is roughly 80°.

Calcite.—The simple carbonate calcite is uncommon in the ores of the Mother Lode belt.

Chalcopyrite.—Chalcopyrite is a very minor constituent of the gold ores of the Mother Lode belt. It was noted to be most abundant in the Melones mine, where it occurs as coarse blebs intergrown with pyrite, with galena, and with tetrahedrite.

Chlorite.—The green micaceous mineral chlorite is common in some of the ores of the Mother Lode mine.

[74] Koller, P., Neues Jahrb., Beilage-Band 42, p. 487, 1918.

[75] Foote, H. W., and Bradley, W. M., On solid solution in minerals: V, The isomorphism between calcite and dolomite: Am. Jour. Sci., 4th ser., vol. 37, pp. 339–345, 1914.

[76] Ford, W. E., op. cit., pp. 229–241.

[77] Turner, H. W., U. S. Geol. Survey Seventeenth Ann. Rept., pt. 1, p. 679, 1896.

Some of it resembles mariposite, some resembles serpentine; and unless optical or chemical tests are made much of it is likely to be identified erroneously. The variety ripidolite was identified as occurring in the Pacific mine, at Placerville. Clinochlore is abundant at Carson Hill.

Damourite.—See Sericite.

Dolomite.—See Ankerite, which is the prevailing carbonate in the gold ores.

Galena.—Galena occurs sparingly in the gold-quartz ores and is generally regarded with high favor, as its presence indicates increased gold content. This correlation of galena and high gold was noticed early in the history of mining in the Mother Lode belt, and that it is a real correlation can be regarded now as well established. Gold is not uncommonly intergrown with the galena.

Gold.—Although the Mother Lode ores are of very moderate grade, gold occurs fairly commonly as easily visible particles and small masses. In the three deepest mines it is sufficiently abundant to make prevention of its theft ("high grading") a matter of practical concern to the managements. The gold may occur alone embedded in quartz or it may be intergrown with pyrite, arsenopyrite, or galena, or rarely with tellurides, generally petzite. It is not everywhere limited to a quartz matrix but may occur also in ankerite and in albite veinlets; nor is it wholly restricted to veins and veinlets but may occur also in the rocks adjacent to them. In the rocks it may be intergrown with sulphides, pyrite most commonly, or it may occur free. A notable occurrence of free gold in country rock has been found at the Clio mine, in Tuolumne County. This gold comes from the 500-foot level, occurs in thin plates, and as it is in unoxidized rock is clearly of primary origin.

The largest piece of gold ever found in California came from Carson Hill, Calaveras County.[78] It was found in November, 1854, weighed 195 pounds troy, or 2,340 troy ounces, and on the assumption that it was 0.900 fine had a value of $43,534. Several other nuggets, weighing from 6 to 7 pounds, were found at the same locality. The large nugget is believed to have been taken out of a quartz vein. According to Lindgren,[79] if it was not directly in a quartz vein, it "was at any rate immediately below the croppings and not in any well-defined alluvial channel."

Apparently there is but one analysis of Mother Lode gold on record.[80] This is an analysis by F. Claudet of quartz gold from the Mariposa estate. There are available, however, a large number of determinations of the fineness of gold bullion. As shown by the data

from the Kennedy mine, the fineness of the gold bullion depends on its metallurgical history, the bullion obtained from the battery amalgam being the purest. There is some evidence that the gold near the surface was of higher chemical fineness than that from greater depth, probably owing to refining of the gold in the zone of oxidation.[81]

On account of the dearth of analyses of gold from the Mother Lode belt four specimens of gold were selected for determination of their fineness. After being thoroughly freed from their associated minerals by Prof. R. K. Warner, they were carefully analyzed by Prof. W. E. Milligan, of Yale University.

Fineness of gold of Mother Lode belt

[W. E. Milligan, analyst]

Locality	Specimen	Fineness
Georgia Slide_____	Crystal of gold from limonitic quartz veinlet.	0.912
Central Eureka mine__	Associated with arsenopyrite__	.839
Kennedy mine, 4,350-foot level.	Associated with pyrite in a quartz gangue.	.825
Melones mine, 1,350-foot level.	Associated with galena in gangue of quartz and dolomite.	.870
Melones mine_____	Associated with petzite in dolomite gangue.	.899

These results are interesting; they suggest that the fineness of the gold is influenced by the nature of the sulphides or tellurides with which it is associated. The exceptional purity of the gold from the Georgia Slide is doubtless due to partial refining in the zone of oxidation.

Gypsum.—In 1914 I found gypsum in the ore on the 2,100-foot level of the Utica mine at Angels. The ore, which carries $2 in gold to the ton, consists chiefly of quartz, with minor dolomite, gypsum, and albite. The only sulphide present is a little galena. The vein is inclosed in amphibolite schist, and the vein and country rock are so impervious that the mine workings were dusty, although the mine above the 900-foot level made large quantities of water. The gypsum occurred in platy aggregates large enough to be recognized underground and was intimately intergrown with the quartz. This intimate intergrowth, together with the occurrence of the gypsum so far below the zone of oxidation and the imperviousness of the vein to descending meteoric water, indicates that the gypsum is a primary (hypogene) constituent of the ore. Under the microscope the gypsum is seen to be intergrown with the quartz in patterns somewhat like micrographic intergrowths, and this feature appears to corroborate the evidence that it is of primary origin.

Magnetite.—Magnetite is disseminated as abundant small well-crystallized octahedrons through a talcose

[78] Hanks, H. G., California State Mineralogist Second Ann. Rept., p. 148, 1882. See also Blake, W. P., The various forms in which gold occurs in nature: Director U. S. Mint Rept. for 1884, pp. 21-22, 1885; in this article he says that the mass included also 4 pounds of quartz.

[79] Lindgren, Waldemar, The Tertiary gravels of the Sierra Nevada of California; U. S. Geol. Survey Prof. Paper 73, p. 66, 1911.

[80] Quoted by Burkart, Neues Jahrb., Jahrg. 1870, p. 173.

[81] Knopf, Adolph, The fineness of gold in the Fairbanks district, Alaska: Econ. Geology, vol. 8, p. 802, 1913.

ankeritized serpentine dike encountered in the Melones tunnel 3,400 feet from its portal. The serpentine had been altered to a highly foliated talc schist, and later ankerite and magnetite were formed in the schist by replacement. This material is undoubtedly the ore from the Melones mine described by Lindgren,[82] who drew attention to the fact that it was the first observed occurrence of magnetite as a product of rock alteration in a California gold-quartz vein, but he described the magnetite as embedded in the "usual sericite felt of the altered amphibolite." This occurrence of magnetite in this talc schist is the only one so far found.

Margarodite.—See Sericite.

Mariposite.—The beautiful green micaceous mineral mariposite is extremely abundant along the southern part of the Mother Lode. It occurs from Carson Hill south to the end of the lode. Notable localities are Carson Hill, the Rawhide mine, Quartz Mountain south of Jamestown, the Mary Harrison mine, near Coulterville, and the Josephine mine, on the Mariposa grant, from which came the original material described by Silliman and the material later analyzed by Hillebrand.

The mariposite occurs as a minor constituent of the immense bodies of ankerite (ferrodolomite) rock that are common along the southern part of the lode. Although a minor constituent, it imparts so distinctive an appearance to the ankerite masses that they are locally spoken of as mariposite rock. This rock has resulted from the hydrothermal alteration of serpentine, and the mariposite is restricted to ankeritized rock derived from serpentine. Locally chlorite and a yellowish sericite are erroneously identified as mariposite. In the Melones mine an emerald-green chlorite occurring in the molybdenitic ore of the 4,125-foot level exactly resembles mariposite. It could be discriminated only under the microscope, where it was seen to be, unlike mariposite, optically positive and feebly birefringent. Mariposite can be identified only by means of chemical and microscopical tests; ordinary visual identification is worthless.

The most conspicuous feature of mariposite is its green color. Emerald-green is most common, but lighter hues occur, such as apple-green, and Hillebrand and Turner speak of a white mariposite, though they do not make clear how such a mariposite is distinguishable from sericite.

Under the microscope the mariposite is seen to occur in finely scaly masses whose optical properties are like those of sericite. It is generally colorless as seen in thin sections, but some show a faint green pleochroism. Only one, from the 2,170-foot level of the Melones mine, gave a distinct pleochroism. The green is deepest parallel to the cleavage.

Mariposites from four localities—the Rawhide mine, the Mary Harrison mine, and two levels of the Melones mine—were carefully examined in oils. Preliminary tests had shown that they all give fine chromium beads, thus conforming to the accepted belief, first put forward by Hillebrand, that the green color of the mariposite is due to the presence of chromium. Thicker flakes all show a fine emerald-green color. The mineral has straight extinction and positive elongation, and it is optically negative and is either practically uniaxial or of narrow axial angle. The mariposite from the Rawhide mine has the widest axial angle, and as measured according to the Mallard method 2 E proved to be 36°. The index γ of three of the mariposites is 1.61, and α, as determined on the mineral from the Mary Harrison mine, is 1.56. The mariposite from the 2,170-foot level of the Melones mine, which unlike the others is distinctly pleochroic in thin section, has notably higher indices: $\gamma = 1.63$ and $\alpha = 1.58$.

Through the courtesy of the United States National Museum I was enabled to examine the mariposite (S. N. 438) collected by Turner and analyzed by Hillebrand. It is nearer apple-green than emerald-green and is associated with a coarse iron-bearing dolomite, an ankerite whose index is 1.70. As seen in refractive index liquids, it is much more weakly colored than the mariposite from the Mary Harrison, Rawhide, and Melones mines; it has not the rich emerald-green that they show in thicker flakes. It is practically uniaxial and its index γ is 1.60. It incloses a few minute prisms of rutile, which account for the titania found by Hillebrand.

Mariposite is clearly a mineral of varied properties, the range in its most distinctive property, its green color, being doubtless due to differences in the quantity of chromium that it contains. Two analyses have been made by Hillebrand on material collected by Turner[83]—one of a green mariposite and the other of a so-called white variety. Both analyses are essentially identical—indeed, within the limits of analytical error—except that the green mariposite contains 0.18 per cent of Cr_2O_3.

Schaller[84] has sought to show the identity of mariposite and alurgite, but the argument is not conclusive, as alurgite is a purple to cochineal-red mica. The mineral mariposite is evidently only of varietal rank; it is a green chromiferous sericite. As the present study indicates, it is not of fixed composition but has a considerable range.

Winchell[85] has recently interpreted mariposite as $Phe_{50}Pro_{50}$—that is, as a solid solution of 50 per cent of phengite ($H_4K_2Al_4Si_8O_{25}$) and 50 per cent of protolithionite ($H_4K_2Fe''_3Al_4Si_5O_{22}$). This particular formula of course would apply only to the mariposite analyzed by Hillebrand.

[82] Lindgren, Waldemar, Metasomatic processes in the gold deposits of Western Australia: Econ. Geology, vol. 1, p. 544, 1906.

[83] Turner, H. W., Further contributions to the geology of the Sierra Nevada: U. S. Geol. Survey Seventeenth Ann. Rept., pt. 1, pp. 678-679, 1896.
[84] Schaller, W. T., The probable identity of mariposite and alurgite: U. S. Geol. Survey Bull. 610, pp. 139-140, 1916.
[85] Winchell, A. N., Studies in the mica group, II: Am. Jour. Sci., 5th ser., vol. 9, pp. 419-430, 1925.

A noteworthy occurrence of mariposite is that at the Croesus copper prospect, in Tuolumne County. (See Pyrrhotite.) The mariposite is here associated with distinctly copper rather than gold deposits.

Molybdenite.—Molybdenite occurs in some abundance in the hanging-wall ore bodies in the 4,125 and 4,250 foot levels of the Melones mine. It is distributed through the highly ankeritized schist, plating the foliation planes in slickensided films. It resembles graphite closely, except for its bluish cast, and the ore in which it is found is locally known as graphitic ore; it is regarded with favor, as indicating more than average gold content in the ore in which it appears. An emerald-green chlorite resembling mariposite occurs with the molybdenite in some of the ore.

It was found to occur also in a pillar of "graphitic" ore just below the 300-foot level of the same mine. As in the bottom levels of the mine it occurs here also in the hanging-wall ore body but in a band of schist that rests immediately against the Bull vein. It is disseminated in flakes so minute as to give a faint blue shimmer to the ore. That it is molybdenite instead of graphite was verified as follows: Some of the rock powder was strongly ignited in a porcelain crucible, a drop of strong sulphuric acid was added and evaporated nearly to dryness; on cooling it showed the rich deep-blue color that, according to Hillebrand,[86] proves the presence of molybdenum.

Molybdenite occurred, according to Ransome, in minor quantity in the ore of the Zeila mine, near Jackson.

Petzite.—The commonest gold-bearing telluride is petzite $(Ag,Au)_2Te$, but it is nowhere abundant. The analyses of petzite from different localities on the Mother Lode belt agree closely in showing 25 per cent of gold and 41 per cent of silver. It is blackish lead-gray, has a conchoidal fracture, and has a brilliant luster on freshly broken surfaces but tarnishes to a peculiar greenish iridescence on exposure. It occurs on Carson Hill embedded in dolomite, where it is associated with abundant free gold, with which it is partly intergrown. It is probably most abundant at the Norwegian mine, on the south side of Stanislaus River. It occurs in some abundance in the Ford mine, at San Andreas, where it is embedded in quartz and is associated with much free gold. It appears to be near hessite (Ag_2Te) in composition.

The other tellurides known from the Mother Lode belt—calaverite, tetradymite, altaite, coloradoite, melonite, and sylvanite—are extremely rare. Most of them came from the Stanislaus mine, on the south flank of Carson Hill.

Prehnite.—Prehnite was discovered microscopically to be fairly abundant in the tuffs and breccias on the west border of the greenstone belt west of Plymouth.

It has been formed along with calcite at the expense of crystals and fragments of albite, which themselves may have been formed as the result of the albitization of calcic plagioclase.

Pyrite.—Pyrite is by far the most abundant sulphide in the gold ores. In the quartz ores it constitutes between 1 and 2 per cent of the ore; in the ores of mineralized country rock, such as the so-called gray ores, it makes up from 2 to 4 per cent of the ore.

The pyrite is in the form of cubes and pyritohedrons, which are generally striated. When the crystals are removed from their matrix, the striations are seen to be impressed on the enveloping quartz. From similar facts observed in other quartz veins Fournet[87] concluded that this proved that the pyrite has crystallized from a quartz magma, which has been injected "instantaneously and completely" into the fissure that the vein now occupies. Others regard this fact as indicating that the pyrite grew suspended in a gelatinous medium which has subsequently crystallized to quartz. As a third possibility the pyrite may have grown by replacement of the quartz. That the third method is by no means improbable is shown by the common occurrence of pyrite in the wall rocks, where much of it has replaced quartz. Probably, however, most of the pyrite in the quartz veins was deposited nearly at the same time with the quartz that surrounded it; possibly some is of replacement origin. The pyrite of known replacement origin—namely, that in the mineralized country rocks—may form large crystals; the largest, about an inch in diameter, appear to be those common in the ore adjacent to the "flat veins" of Carson Hill.

Gold intergrown with pyrite was detected microscopically, as in the mineralized amphibolite schist from the Gold Cliff mine, where it was of contemporaneous origin; doubtless more intensive search would reveal it to be common as extremely minute particles embedded in pyrite.

Pyrrhotite.—Pyrrhotite was found at only one typical Mother Lode mine—the Treasure mine, in Amador County. It occurs there in albite veinlets that ramify through augite melaphyre; it is highly magnetic. It occurs abundantly in the ore at the Croesus prospect, on Merced River 2 miles north of Bagby, where it is associated with chalcopyrite and pyrite in ankeritized serpentine, some of which contains mariposite. As the ore carries gold also, it is clear that this essentially copper mineralization has a close affinity with that of the Mother Lode. Pyrrhotite is disseminated through the richly hornblendic diorite at the Easy Bird mine, northeast of Mokelumne Hill, a few miles east of the Mother Lode belt.

Quartz.—The quartz of the Mother Lode veins is a coarse milk-white variety, of a luster between glassy

[86] Hillebrand, W. F., The analysis of silicate and carbonate rocks: U. S. Geol. Survey Bull. 422, p. 151, 1910.

[87] Fournet, J., Sur l'état de surfusion du quartz dans les roches éruptives et dans les filons métallifères: Compt. Rend., vol. 18, pp. 1050–1057, 1844.

and greasy. Microscopically, where undeformed by pressure after its deposition, the quartz ore is seen to have a hypidiomorphic granular texture, being composed of large optically homogeneous units that have in part rectilinear boundaries. The quartz contains innumerable inclusions, probably liquid, so minute as to be hardly more than points even under high-power objectives.

Much of the quartz has been crushed; in places it has been rendered sugary. Microscopically the quartz shows all stages of cataclastic deformation, from strain shadows to granulation so complete that the larger quartz grains have been broken into hundreds of smaller grains. This feature is shown in a particularly striking way by good-grade ore from the 4,200-foot level of the Argonaut mine. This ore was inclosed by thick gouges on both foot and hanging walls. To the eye the vein was obviously the result of filling at intervals, as shown in Plate 3, A; it also suggests that the quartz is partly of replacement origin on account of the innumerable filaments of black slate inclosed in it, which give the impression that they are residuals from incomplete replacement. The microscope, however, does not support this idea. The quartz gives no evidence that it is of replacement origin, and the slate filaments that are inclosed in it are sharply defined and are sericitized, not silicified. The quartz has been greatly crushed, almost completely granulated, but the larger grains that have survived show undulatory extinction and a well-defined narrow striping, which is emphasized by the fact that the minute liquid(?) inclusions already mentioned are markedly localized along a series of parallel cracks. The quartz was evidently sliced along a fairly regular system of parallel fractures, and the inclusions were deposited along or adjacent to them. This feature is identical with that described and illustrated by Mügge.[88]

The quartz of the veins not uncommonly contains small vugs, and well-defined faces of coarse quartz crystals form the sides of these vugs or the crystals project into the vugs. This habit of the quartz appears to be an insuperable argument against the idea that the quartz has crystallized from a silica gel, for a silica gel on crystallizing yields, so far as we know now, only radially fibrous chalcedony or cryptocrystalline quartz, not coarsely crystalline quartz.

It is an interesting fact that much of the quartz in the gold veins east of the Mother Lode belt is perceptibly different from the milk-white variety of the Mother Lode belt. This feature was observed in the region east of San Andreas, where the quartz has a peculiar smoky-gray greasy aspect; the most notable example is at the Sheep Ranch mine. At the McGary mine, 3 miles south of Railroad Flat, although most of the quartz is milky white, yet that containing pyrite, galena, and sphalerite is smoky.

The microscopic features of the vein quartz were first investigated by Phillips,[89] who was chiefly interested in the liquid inclusions contained in the quartz. When a section of a crystal of quartz from the great Mariposa vein was heated, the vacuities in six fluid cavities disappeared at 250°, 260°, 280°, 290°, 310°, and 320° F. Others were not nearly full at 362° F. From these determinations Phillips[90] drew the conclusions that the fluid cavities are the result of an intermittent action, and that the fissures were sometimes traversed by currents of hot water, whereas at others they gave off aqueous vapor or gaseous exhalations. Just what conclusions can now be drawn are doubtful.[91]

Courtis[92] thought that there is some evidence that rich gold quartz is crowded with fluid inclusions arranged in parallel bands. It is known that the number of inclusions in a mineral varies directly as the rate of crystallization, and their size varies inversely,[93] but that there is any relation between the number of liquid inclusions and the amount of gold is not established.

Riebeckite.—Riebeckite occurs abundantly in tufts and radial groups of needles in the albitite near the Clio mine. It has doubtless been formed at the expense of aegirine, which is the normal constituent of the albitite. Its distinguishing features are its striking pleochroism, ranging from green to deep blue; negative elongation; extinction of 5°; and feeble birefringence, about 0.004.

Rutile.—Rutile is a microscopic constituent of the ankeritic schist ore of Carson Hill. It occurs both in the high-grade ore of the hanging-wall ore body and in the low-grade material formerly extensively mined in the footwall of the Bull vein; it is present as small prisms and geniculate twins. Its occurrence is reminiscent of its abundance in the gold ores of the Juneau belt, Alaska. In the mariposite of the Mary Harrison mine it is inclosed as minute prisms and as heart-shaped and geniculate twins.

Sericite.—The finely scaly variety of muscovite is extremely common in the altered wall rocks of the gold deposits. Its usual associates are ankerite, albite, and quartz. It varies greatly in appearance from place to place, in part doubtless as a result of or in response to differences in the original nature of the wall-rock material and in part in response to differences

---

[88] Mügge, O., Neues Jahrb., Beilage-Band 10, p. 767, pl. 14, fig. 2, 1896.

[89] Phillips, J. A., Notes on the chemical geology of the gold fields of California: Philos. Mag., vol. 36, p. 333, 1868; abstract in Am. Jour. Sci., 2d ser., vol. 47, pp. 134–140, 1869.

[90] Idem, p. 432.

[91] Nacken, R., Welche Folgerungen ergeben sich aus dem Auftreten von Flüssigkeitseinschlüssen in Mineralien: Centralbl. Mineralogie, Jahrg. 1921, pp. 12–20, 35–43.

[92] Courtis, W. M., Gold quartz: Am. Inst. Min. Eng. Trans., vol. 18, pp. 639–644, 1890.

[93] Rosenbusch, H., Mikroskopische Physiographie der Mineralien und Gesteine, Band 1, Erste Hälfte, p. 784, 1924.

in the composition and temperature of the ore-forming solutions. Under the microscope, however, all these sericites look alike. A practical difficulty that arises in the study of Mother Lode rocks is the impossibility of distinguishing talc optically from sericite. Both occur in great abundance and in places together in the same rocks. In all early reports the sericitic rocks are invariably referred to as talcose.

White, yellowish, and greenish tints are common. At Carson Hill, for example, the sericite in the mineralized schist that constitutes the ore bodies is of a peculiar oily, yellowish appearance. The coarse white quartz ore in the bottom levels (4,200 feet vertical depth) of the Kennedy mine, as seen underground, looks in places as if it had been daubed with aluminum paint. This appearance is due to a silvery or pearly mineral in crumpled talclike flakes, which is locally called pyrophyllite but in reality is the variety of muscovite allied to damourite. The scales, which are not elastic, are large enough to be examined microscopically. The cleavage flakes prove to be isotropic and give a negative uniaxial interference figure; $\gamma = 1.59$. In addition the mineral reacts for potassium and aluminum and is therefore a variety of muscovite. Evidently it is the mineral referred to by Ransome in the Mother Lode folio as sericite or margarodite. Margarodite is regarded by Dana as a synonym for damourite.

Similar pearly mica occurs in the quartz of the Keystone mine, where it was correctly identified by Prichard [94] as margarodite. It occurs abundantly also in the so-called flat veins in the Melones mine, on Carson Hill. In the Central Eureka mine sericite occurs in the quartz in places as compact masses of almost waxlike appearance. However, there are from place to place all gradations from the waxlike to the coarse scaly talclike variety, damourite.

Sphalerite.—Sphalerite (zinc blende) is a common minor constituent of the gold ores. Apparently it is restricted to the quartz of the veins and does not occur in the adjacent wall rocks, being in this respect unlike pyrite and arsenopyrite, which are common as replacement deposits in the wall rocks. The sphalerite ranges in color from resinous through brown to black. It extends to the greatest depth attained, being found in small quantity in the bottom levels of the Kennedy mine.

Talc.—Talc is common in some of the schists, especially those derived from peridotites or their serpentinized equivalents. Talc is not easily distinguishable from sericite, except that it is softer and marks cloth. Optically the two minerals are indistinguishable, though all the talc that was examined proved to be practically uniaxial, whereas the sericite as a rule has a moderately wide axial angle—2E

about 40°. This test can rarely be applied, and when it can be it is not conclusive, for sericite, especially the variety damourite, previously described, and mariposite may have axial angles near 0°. The flame test for potassium must be used in critical work.

Tetrahedrite.—Tetrahedrite (gray copper) occurs at Carson Hill and in Mariposa County. It was seen during the present survey in ore from the 1,100-flat vein of the Melones mine, Carson Hill, where it formed coarse blebs intergrown with galena and chalcopyrite in ankerite-quartz veinlets. It was seen also as a very minor constituent in the quartz ore of the new Princeton vein, near Mariposa.

ALTERATION OF WALL ROCKS BY THE ORE-FORMING SOLUTIONS

ALTERATION OF THE MARIPOSA SLATE

The black slates adjacent to the veins have obviously been altered, as is made manifest by the small lenticular grains of carbonate that have formed in them. These lenticles or augen of carbonate cause the cleavage surfaces of the slates to be dotted with small depressions or pits and with the corresponding swellings or pustules. Slates have been altered in this way as far as 10 feet from the veins, and as the slates are essentially parallel to the veins the alteration has visibly affected a belt at least 10 feet thick. This alteration is not dependent on fissuring of the slate but was clearly the result of thorough permeation of the adjacent slates by the ore-forming solutions.

Under the microscope the carbonate augen are seen to be mainly ankerite, as indicated by the occasional twinning bands that form the shorter diagonal of the cleavage rhombs. Possibly some are breunnerite, $(Mg,Fe)CO_3$, as indicated by the chemical analysis on page 42. They show a remarkable feature, apparently never before noticed in the study of the altered wall rocks of ore deposits but commonly seen in the study of the crystalline schists—namely, that many metacrysts or porphyroblasts in schists have been partly rotated during their growth by the continued "flow" of their embedding matrix.[95] The carbonate augen, although in optically uniform orientation over their entire areas, faithfully reproduce the original layered structure of the slate. This pseudomorphic replacement of the slate is well shown in Plate 7, *A*, *B*, and *C*. The carbon in the slate was not carried away during the growth of the carbonate augen; consequently all the augen are dark, and some are nearly black. The layering within the augen is askew to that in the unreplaced slate; in some of the augen it differs as much as 25°; moreover, the layering of the slates tends to wrap around the carbonate augen. At first sight the circumfluent foliation would seem to

[94] Prichard, W. A., Observations on Mother Lode gold deposits, California: Am. Inst. Min. Eng. Trans., vol. 34, p. 458, 1903.

[95] Becke, F., Struktur und Klüftung: Fortschritte der Mineralogie, Kristallographie und Petrographie, Band 9, pp. 194–199, 1924.

indicate that the growing metasomes of ankerite had crowded aside the slate by its force of crystallization, but the real explanation appears to be that movement took place after the metasomes were formed and that differential shear parallel to the bedding of slate caused them to be rotated through angles as great as 25°. Many of the carbon-pigmented augen are surrounded by fringes of perfectly clear unpigmented carbonate, as is shown in Plate 8, A. This clear carbonate might be interpreted as the filling of open spaces developed during the partial rotation of the pigmented augen—that is, carbonate deposited in the "pressure-shadows" of the augen—but a more likely explanation is that it also has been formed by replacement of the slate, except that at this stage of replacement the solutions had become capable of dissolving the carbon also and carrying it away. That this is the more probable explanation is shown by the sporadic occurrence of sharp idiomorphic rhombs of clear carbonate within the pigmented augen (see pl. 7, B) and by the occurrence of metasomes of at least two generations in some slates, the earlier highly pigmented and partly rotated and the later clear, spongiform, with the interspaces of the "sponges" filled with albite. The carbonate of later origin may occur as sharp rhombs that cut cleanly across the cleavage of the slate. Obviously all or nearly all the material of the slate, including the carbonaceous matter, was carried away in the process of forming the clear ankerite.

Rarely the ankerite metasomes show an S-shaped deformation of the inclosed slate laminae; from the amount of deformation of this kind Becke has shown that a minimum measure of the flowage of the slate during the growth of the metasome can be computed. A few metasomes became cracked during the flowage of the matrices in which they lie and were later healed by the deposition of more carbonate.

In addition to the metasomatic carbonate sericite has formed, as the result either of the introduction of new material or of recrystallization or both. At any rate it is in remarkable alinement, so that the slate extinguishes with the sharpness of a single large crystal. Metasomatic pyrite is common in sharp crystals, which invariably transect the cleavage of the slate—a feature that points to the relatively late introduction of the pyrite, after flowage of the slate had ceased. Associated with some of the larger pyrite crystals is the quartz with the curious structure that causes it to be termed feather quartz. (See pl. 8, B.) At only one locality, the Sherman mine, near Placerville, was any metasomatic tourmaline seen, and there it was only in minute quantity, associated with coarse albite, quartz, and ankerite, which form the cement of a slate breccia.

The predominant metasomatic minerals thus found by the microscope are carbonate and pyrite. Many thin sections of slates adjacent to veins and occurring as inclusions in the veins were examined, and all concur in showing that carbonatization was overwhelmingly the predominant process. No support is given to the idea that the quartz veins are the results of replacement of slate by quartz. The following chemical analysis indicates that considerable albite was also formed. It is of course impossible to obtain slate that can be demonstrated to be the unaltered equivalent of the altered slate. In order to indicate in a general way the trend of the changes produced in the slate by the ore-forming solutions, the only two available analyses of Mariposa slate are given in the subjoined table. Analysis 2, it is apparent, is that of a normal marine shale or slate free from admixed carbonate.

Analyses of altered and normal Mariposa slate

	1	2	3
SiO_2	57. 21	60. 35	63. 52
Al_2O_3	11. 76	17. 62	16. 34
Fe_2O_3	1. 22	5. 64	} 6. 79
FeO	4. 30	2. 20	
MgO	3. 58	1. 04	2. 50
CaO	3. 43	. 45	. 98
Na_2O	2. 84	1. 00	------
K_2O	2. 47	3. 16	------
H_2O-	. 06	1. 02	} [a] 4. 86
H_2O+	. 83	4. 36	
TiO_2	. 70	. 75	------
P_2O_5	Not det.	. 17	------
BaO	Not det.	. 12	------
CO_2	8. 37	None.	------
SO_3	Not det.	. 05	------
FeS_2	2. 14	------	------
Carbonaceous matter	1. 50	1. 72	------
	100. 76	99. 71	------

[a] Includes CO_2.

1. Slate from hanging wall of Argonaut vein, in shaft crosscut, 4,650-foot level. J. G. Fairchild, analyst.
2. "Clay" slate near head of Yaqui Gulch, Mariposa County. George Steiger, analyst. U. S. Geol. Survey Bull. 591, p. 257, 1915.
3. Slate from Eureka quarry, Slatington, Eldorado County. W. T. Schaller, analyst. U. S. Geol. Survey Bull. 586, p. 67, 1914.

The conclusion appears warranted that silica, alumina, and water have been removed from the slate, that the ferric iron has been largely reduced to the ferrous state, and that carbon dioxide and sulphur have been added in notable quantities, as well as calcium, magnesium, and sodium.

The mineral composition of the altered slate from the Argonaut mine is computed to be as follows, on the assumption that the sericite corresponds to the formula $K_2O.2H_2O.3Al_2O_3.6SiO_2$:

Quartz	31. 3
Albite	23. 6
Carbonate	20. 6
Sericite	21. 8
Pyrite	2. 1
Carbonaceous matter	1. 5
	100. 9

ALTERATION OF GRAYWACKE, CONGLOMERATE, AND QUARTZITE

Graywackes and conglomerates of the Mariposa formation and quartzites of the Calaveras formation have been metasomatically altered near the veins, and all these rocks show that the predominant effect has been ankeritization, with subordinate albitization, sericitization, pyritization, and introduction of arsenopyrite. The ankerite is abundant, is generally in sharp rhombs, and has developed indifferently through all the constituents of the rock, including the quartz. Much of it shows twinning bands, which bisect the obtuse angles of the rhombs. Some rhombs show a zonal structure. In places the quartz grains in the graywacke had been crushed, evidently by the compressional forces that produced the fissuring, and were later replaced by ankerite.

The fine-grained quartzite at the Eagle-Shawmut mine has taken on an easily recognizable pseudoporphyritic texture as a result of the formation through it of euhedral rhombs of ankerite. Albite occurs in places as large twinned grains associated with the carbonate, and it occurs also as a fine-grained replacement product of the quartz.

ALTERATION OF THE GREENSTONES

The greenstones have been intensely altered by the ore-forming solutions. Ankerite and sericite are the chief minerals that were thus formed. Near the veins the greenstones are commonly solid ankerite rocks, and as the ankeritization has affected large volumes of rock it is clear that enormous quantities of silica have been removed from the wall rocks.

Sericite is the next most abundant constituent and is followed by albite, quartz, pyrite, and arsenopyrite. The sulphides are invariably sharply euhedral. In places the ankeritic sericitic rocks contain gold, and constitute the so-called gray ore. Chlorite occurs where the alteration has been feebler.

The greenstones preserve their original structure and texture fairly well, in spite of the drastic alteration to which they have been subjected; locally, in fact, the original pyroclastic breccia structure has been emphasized by the selective alteration of the smaller fragments that form the matrix in which the larger pieces are embedded. Under the microscope many of the fully ankeritized augite crystals in the greenstones are found still to retain their original zonal structure.

Representative gray ore from the 1,750-foot level of the Bunker Hill mine was selected for chemical analysis. It is a gray rock containing a moderate quantity of euhedral pyrite and some arsenopyrite, also in well-formed crystals. It does not effervesce with hydrochloric acid. Microscopically it is seen to be composed mainly of ankerite and sericite. The carbonate ($\omega = 1.705$) is largely in irregular spongiform patches, and the sericite (2 E estimated to be 40°) is generally idiomorphic against it. Quartz and albite

are subordinate constituents; the albite is difficult to recognize, and in fact its presence was rendered certain only by testing some of the crushed ore in index liquids. Rutile occurs in minute crystals, and possibly some leucoxene is present. The following chemical analysis confirms the microscopic analysis, although the content of carbon dioxide appears to be somewhat low. The analysis of the augite melaphyre described on page 15 is given also, in order to indicate in a very general way the trend of the alterations that have affected the greenstone.

Analyses of hydrothermally altered greenstone and augite melaphyre from Mother Lode belt

	1	2		1	2
SiO$_2$	34. 17	49. 24	H$_2$O+	1. 94	2. 97
Al$_2$O$_3$	15. 82	14. 79	TiO$_2$	1. 49	. 96
Fe$_2$O$_3$	. 69	1. 36	P$_2$O$_5$	Not det.	. 17
FeO	7. 61	8. 00	MnO	Not det.	. 18
MgO	a 5. 77	6. 89	BaO	Not det.	. 04
CaO	8. 75	10. 74	CO$_2$	15. 50	. 90
Na$_2$O	1. 65	2. 76	FeS$_2$	2. 74	--------
K$_2$O	4. 63	. 88			
H$_2$O−	. 00	. 20		100. 41	100. 08

a Some MgO probably is combined with SiO$_2$.

1. Hydrothermally altered greenstone ("gray ore") from Bunker Hill mine. J. G. Fairchild, analyst.
2. Augite melaphyre from point west of Jackson. W. F. Hillebrand, analyst.

The main changes brought about in the greenstone are the notable removal of silica and the large addition of carbon dioxide and potassium. These changes are clearly apparent from the microscopic examinations, which give a more nearly average picture of the alteration than a single chemical analysis and in fact prove that at many places the greenstones have been altered completely to carbonate.

The analysis is rather difficult to compute into mineral composition. The amount of carbonate obtained by the calculation appears to be somewhat low, the sericite and albite somewhat high. Part of the difficulty is doubtless due to the fact that the formula used in computing the amount of sericite ($K_2O.2H_2O.3Al_2O_3.6H_2O$) is not strictly applicable, as the sericite in the rock probably contains magnesium and other constituents. The computation results as follows:

Ankerite	35. 46
Sericite	39. 90
Albite	13. 62
Quartz	7. 14
Pyrite	2. 74
Rutile	1. 49
	100. 35

ALTERATION OF THE AMPHIBOLITE AND ALLIED SCHISTS

The alteration of the amphibolite and allied green schists is closely similar to that of the greenstones, in that the altered schists are composed chiefly of ankerite and sericite, with subordinate quartz, albite, and pyrite. Rutile is a common accessory, forming minute prisms and geniculate twins. As the foliated texture of the schists survives through the alteration, or possibly is even emphasized by the newly formed sericite, the actual appearance of the mineralized schists differs greatly from that of the hydrothermally altered greenstones.

Pyrite is developed metasomatically in the schists as sharp cubes cutting the foliation; in places, as at the Gold Cliff mine and Carson Hill, the cubes attain large size—half an inch or more in diameter. Gold is occasionally found inclosed in the pyrite. Molybdenite is the only other sulphide occurring by metasomatic replacement of these schists.

ALTERATION OF SERPENTINE

The serpentine and serpentinized peridotites have been more extensively altered by the vein-forming solutions than any other rocks. They have been converted into ankerite rocks containing more or less mariposite—a green chromiferous sericite. (See p. 38.) As mariposite does not occur in any rocks not dereived from serpentine, it is probable that the chromite content of the serpentine was essential to the forming of the mariposite. Immense masses of such mariposite-ankerite rocks occur from Angels Camp southward to the end of the lode.

The ankerite is generally coarse, but the crude foliated structure of the serpentine is faithfully preserved. On oxidation the ankerite yields a limonitic gossan, as is well shown, for example, at the Mary Harrison mine, south of Coulterville.

The mariposite-ankerite masses are generally traversed by a network of veinlets, consisting of coarse milky-white quartz. In places where they are closely spaced and the intervening rock has been more r less replaced by quartz huge masses of quartz have resulted, like those forming the great outcrops at the Eagle Shawmut mine, shown in Plate 5, B. The quartz of these masses is clearly of composite origin, yet it is all of the coarse white variety commonly considered to be formed only by deposition in open spaces.

Probably one of the most interesting alterations of serpentine is that seen on the 1,350-foot level of the Melones mine. Where unaltered by the vein-forming solutions the serpentine is a rather massive, finely phanerocrystalline variety composed of antigorite. It has been sericitized, and all stages of replacement of the antigorite can be followed. Some carbonate and pyrite were also formed contemporaneously with the sericite. On account of the difficulty of distinguishing sericite from talc, the identity of the sericite was established by the potassium flame reaction in hydrofluoric acid solution. As already pointed out (p. 20), this sericitized serpentine proves that the peridotite or pyroxenite from which it was derived had already been altered to serpentine at or near the end of Jurassic time, or in other words before the end of the petrogenic cycle in which the peridotite originated.

ALTERATION OF CHLORITE SCHIST (SHEARED PYROXENITE)

The remarkable "sulphide ore body" at the Eagle Shawmut mine, the only one of its kind so far found in the Mother Lode belt, has resulted from the replacement of a blackish-green compact chlorite schist. Megascopically this chlorite schist closely resembles a serpentine, but microscopically it is seen to be composed of an ultra-fine scaly chlorite. As shown on page 79, it is derived from a pyroxenite. The ore body has been formed by the replacement of the chlorite schist by ankerite, pyrite, and arsenopyrite.

SUMMARY OF WALL-ROCK ALTERATION

Carbonatization (ankeritization) has been the dominant process by which the wall rocks were altered, regardless of whether these rocks were slate, graywacke, quartzite, conglomerate, greenstone, amphibolite schist, chlorite schist, talc schist, or serpentine. It has affected large volumes of rock. Second in importance has been sericitization, so that the altered wall rocks can be called sericite-ankerite rocks. In the alteration of the serpentines sericite was formed in much less quantity, but the distinctive chromiferous mica mariposite was developed. Of all the rocks altered, the serpentines were most susceptible to ankeritization, and belts of ankeritized serpentine that are several hundred feet wide at a maximum have been formed. The augitic greenstones were next to the serpentine in sensitiveness to ankeritization.

That carbonatization of the wall rocks was probably the most common effect of the vein-forming solutions throughout the Sierra Nevada was pointed out by Lindgren [96] in his study of the Ophir district, in Placer County. His later investigation of the Grass Valley and Nevada City districts showed the same thing to be true in those districts; [97] and the present investigation has shown the overwhelming importance of this alteration in the Mother Lode belt. In all respects the alterations of wall rock in the several districts are remarkably similar, the chief characteristics being the addition of carbon dioxide, potassium, and sulphur and the removal of silica. The amount of carbon dioxide added to any particular wall rock appears to have been largely determined by the amount of iron, magnesium, and calcium originally contained in the wall rock.

Arsenic has been added in places in notable quantities to the wall rocks and has formed metasomatic arsenopyrite, invariably as sharply defined crystals. Sodium has also been introduced and has formed albite, but the albite has developed regardless of the original chemical nature of the wall rock. No rules can be formulated, and the appeal to "mass action" that has sometimes been made is futile.

[96] Lindgren, Waldemar, The gold-silver veins of Ophir, Calif.: U. S. Geol. Survey Fourteenth Ann. Rept., pt. 2, p. 278, 1894.
[97] Lindgren, Waldemar, The gold quartz veins of Nevada City and Grass Valley districts, Calif.: U. S. Geol. Survey Seventeenth Ann. Rept., pt. 2, pp. 146-157, 1896.

The most novel feature that the study of wall-rock alteration has disclosed is in the slates. It is plainly shown that the ankerite metasomes have been partly rotated at some time during the period when the slate wall rocks were being altered by the vein-forming solutions.

Clearly the most remarkable feature of the wall-rock alteration in the Mother Lode belt has been the displacement of silica by carbon dioxide. Enormous quantities of silica have been liberated from the silicates in which it was combined and carried out of the wall rocks into the fissures, where some of it was deposited and the remainder carried away. The amount of quartz in the veins is only a fraction of the amount of silica liberated from the wall rocks by ankeritization. The wall rocks were a more than ample source of silica for the quartz in the veins, hence it is unnecessary to appeal to a magma as the source of the silica.

The suggestion often made that the wall rocks acted as a semipermeable membrane that prevented the silica, postulated to be in the colloidal state, from moving into the wall rocks is not a particularly happy one, for the movement of silica was all in the other direction—from the wall rocks to the fissures. Had the wall rock been a semipermeable membrane it would have prevented movement of the silica, if colloidal, in either direction. Not only were the ferromagnesian minerals decomposed by the ore-forming solutions, but even the free quartz in the quartzites and graywackes was replaced by ankerite. In short, the solutions that penetrated the wall rocks were avid for silica, free or combined.

The quartz veins in places contain fragments of the wall rocks. These inclusions consist of sharply angular pieces of greenstone, slate, and quartzite. They show identically the same alterations as the wall rocks; no unattacked inclusions were found. As a result of the detailed petrographic examination of many thin sections cut from the inclusions it can be said that they afford not the faintest support to the idea that the quartz veins have been formed by replacement.

The conclusions that flow from the evidence of replacement in the Mother Lode belt are like those drawn by Lindgren from the evidence at Grass Valley. After pointing out the prevalence of replacement at Grass Valley and stating that all the minerals there are subject to it, even the quartz of the granodiorites, he says:[98]

Replacement by silica is not among the processes here recognized. It should be borne in mind that a rock shattered and filled with quartz seams is not an evidence of metasomatic replacement by quartz, nor is such a rock a quartz vein in process of formation.

Howe[99] also was unable in his recent restudy of the Grass Valley district to find metasomatic quartz "as one of the minerals replacing the granodiorite or diabase" (the wall rocks of the quartz veins in that district). Nevertheless, he reaches the conclusion that "the veins were formed by the replacement of wall rock by quartz, calcite, and metallic sulphides."[1] It is difficult to reconcile this conclusion with the absence of metasomatic quartz in the district.

The status of the same problem on the Mother Lode is that all the positive evidence from petrographic studies of wall-rock alteration is adverse to the idea that the quartz in the veins is of replacement origin. The most striking ore body in the Mother Lode belt that is certainly of replacement origin—namely, the very notable "sulphide ore body" at the Eagle Shawmut mine—consists chiefly of ankerite, with pyrite, arsenopyrite, and unreplaced residuals of chlorite schist, but no quartz.

ORIGIN OF THE GOLD DEPOSITS

The Mother Lode veins, as now long recognized and as amply confirmed by the present investigation, occupy fissures that were formed by reverse faulting—that is, by faulting in which the hanging wall has moved up relatively to the footwall. It is interesting to recall, however, that in 1913 the Kennedy Extension Gold Mining Co. attempted to wrest the title to the Argonaut vein from the Argonaut Mining Co., which had been operating 20 years on the vein, on the plea that the Mother Lode vein occupies a normal fault. The court rejected the contentions of the experts on both sides and confirmed the title of the Argonaut Mining Co. to the vein, on the ground that it is manifestly unjust to take away a property, on the basis of a geologic theory, after 20 years of undisturbed possession.[2]

The faults dip less steeply than the slates. The powerful compressive force that produced the faulting has flattened the slates against the veins and produced a belt of schistose material along the fault zone. Naturally, then, in any exposure in a drift on a given vein the vein appears to lie parallel to the structure of the inclosing country rock, and for this reason few of the operators on the Mother Lode realize that the veins as they are followed downward gradually cut through belts of rock of diverse character.

The displacement along some of the faults amounts to 375 feet, but this total displacement is the cumulative result of a considerable number of displacements that occurred over a considerable span of time. That the present structure of the veins proves that they were opened intermittently was recognized by Ransome.[3] The observational evidence for this conclusion lies in the ribbon structure of the veins and the quartz veinlets cutting older quartz. That the compressive forces

[98] Op. cit., Seventeenth Ann. Rept., pt. 2, p. 147.

[99] Howe, Ernest, The gold ores of Grass Valley, Calif.: Econ. Geology, vol. 19, p. 615, 1924.

[1] Idem, p. 619.

[2] Min. and Sci. Press, vol. 109, pp. 61–64, 1914.

[3] Ransome, F. L., U. S. Geol. Survey Geol. Atlas, Mother Lode District folio (No. 63), pp. 7–8, 1900.

that formed the fissures acted during a long span of time, while the vein filling was being deposited, is proved by the rotation of the ankerite augen in the slate wall rocks.

From the fact that the dislocations produced during historic earthquakes are not known to exceed 49 feet and as a rule are much smaller Högbom[4] concluded that faults of much larger magnitude than 49 feet are the results of successive movements. This conclusion appears to be verified geologically by the fact that fault breccias of the larger faults show that they have been repeatedly crushed and recemented. Incidentally, it should be mentioned that Högbom[5] believes that the completely isolated fragments of country rock that occur in the cement of fault breccias have become thus isolated as a result of repeated brecciation and recementation. Flett[6] found that the filling of the Cornish tin lodes, which according to geologic conceptions must have formed within a very short interval of time, shows abundant evidence of repeated crushing and recementation. Moreover, the same conclusion—namely, that a large displacement is the summation of a considerable number of minor displacements—follows as a logical consequence of the elastic rebound theory of faulting developed by Lawson[7] and by Reid[8] from their studies of the San Andreas fault of California. According to this theory faults are due to an elastic rebound on a rupture plane in the crust on which strain is suddenly relieved. The strain is slowly generated during a period of years, and when the strength of the rocks or the frictional resistance to movement on an old rupture is exceeded the rocks rupture. At the time of rupture the displacement is limited to the amount necessary to relieve the accumulated strain. The displacement is therefore proportional to the length of the fault along which the strain accumulates, and to judge from the small amount of elastic distortion that rocks can undergo before they rupture the displacement must necessarily be small—a conclusion that harmonizes with the relatively small displacements known to occur at the time of historic earthquakes.

The great linear extent of the Mother Lode system appears less remarkable since the results of the investigations of the San Andreas fault of California have become known. Here is a fault that has been traced for 600 miles, that is known positively to have been ruptured for a distance of 190 miles and probably 270 miles in 1906, and that traverses rocks of all degrees of strength, from granite to weak sedimentary rocks.

The position of the San Andreas fault manifestly was not determined by a belt of weak rocks but by a deeper-seated, fundamental cause. Auxiliary fractures occur in a zone adjacent to the San Andreas fault.[9]

The location of the Mother Lode fissure system, like that of the San Andreas fault, appears to have been determined by a deeper-seated, fundamental cause than a weak belt of rocks. It traverses rocks as diverse as slate, greenstone, amphibolite schist, and serpentine. In places there is evidence that the veins occupy auxiliary fractures in a zone parallel to a great reverse fault. This evidence is best shown near Plymouth, where amphibolite schist has been thrust up over Mariposa slate. (See figs. 3 and 7.) Some mineralization has taken place along the master fault, but apparently it is nowhere of major importance. At Jackson the structural conditions appear to be like those at Plymouth, whereby the amphibolite schist has been thrust up over the Mariposa rocks. The fault zone is marked by a wide silicified belt, as at Jackson Gate, and 1½ miles farther south an important mine, the Zeila, was located on it. At the Eagle Shawmut mine, as at the Zeila, the main ore bodies were formed in or closely adjacent to the main reverse fault. (See fig. 24 and pl. 12.)

The movement along the fissures, which are markedly sinuous, would, according to the time-honored explanation of Werner, produce open cavities. It is argued by some that cavities would not stay open in slates, but drifts and crosscuts driven through normal Mariposa slate that has not been reduced to gouge remain open indefinitely. However, if the open cavity were large, some adjustment by gravity faulting would most likely take place. This possibility is strongly suggested by what is happening throughout the Argonaut mine as the result of stoping out large shoots of ore. Drill holes show normal faulting; horizontal fissures as much as 3 inches across have opened along joint planes; and highly suggestive of the way in which the ribbon structure of the veins was formed, some of the fissures have opened along horizontal quartz veinlets, the quartz with more or less black slate adhering to its upper surface forming the footwall of the open fissure. Granted that an open cavity existed before the main vein was filled with quartz, subsidence and the resulting opening of joints accounts perfectly for the horizontal quartz veins that in many places cut across the slates in the stringer halo of the main vein.

In recent years many explanations have been advanced to account for the quartz filling of veins of the Mother Lode type. The idea advocated by Lindgren[10] that the clean quartz in a vein is the result of the filling of cavities has proved unacceptable to many. Injection of a quartz magma in the form of a dike is

[4] Högbom, A. G., Zur Mechanik der Spaltenverwerfungen; eine Studie über mittelschwedische Verwerfungsbreccien: Geol. Inst. Upsala Bull., vol. 13, pp. 391–408, 1916.

[5] Idem, p. 398.

[6] Flett, J. S., Notes on some brecciated stanniferous veinstones from Cornwall: Geol. Survey England and Wales Summary of Progress, 1902, pp. 154–159, 1903.

[7] Lawson, A. C., Report of the California Earthquake Commission, vol. 1, pt. 1, pp. 147–151, 1908.

[8] Reid, H. F., Elastic rebound theory of earthquakes: California Univ. Dept. Geology Bull., vol. 6, pp. 413–444, 1911.

[9] Lawson, A. C., op. cit., pp. 53–57.

[10] Lindgren, Waldemar, Characteristic features of California gold quartz veins: Geol. Soc. America Bull., vol. 6, p. 229, 1895.

an idea that has been revived, replacement is an alternative idea, and the pushing apart of the walls of a fissure by the force of crystallization of the growing quartz is still another. It has also been suggested that the original filling of the fissure, especially the gouge, may have been more nearly removed where the ascending currents of the ore-forming solutions were flowing most swiftly.[11] Some, like Tronquoy[12] in explaining the tin-bearing quartz veins of Villeder, France, have gone so far as to maintain that the quartz was injected as a jelly and then slowly became coarsely crystalline. Although this hypothesis appears to account satisfactorily for the "floating" or isolated "unsupported" inclusions of wall rock in the veins, yet it seems not applicable to the Mother Lode veins, because nowhere do they show any colloform structures; vugs are common, but they invariably have sharply angular surfaces (see p. 40) instead of the cauliflower-like surfaces common in gel minerals that have become crystalline. Irregular shrinkage cavities might be produced by the loss of water during the crystallization of the gel, but they would be lined with chalcedony or fine-grained quartz and not with large crystals of quartz.

The hypothesis that the quartz veins are frozen magmatic injections—dikes, in short, or vein-dikes, as Spurr[13] calls them—is believed to be improbable because, first, the quartz veins nowhere show any evidence of the chilling in the form of fine-grained border facies, even if only narrow, that should be found in dikes that were injected into rocks relatively so cold as those of the Mother Lode belt; and, second, the enormous amount of replacement effected in the wall rocks is out of all proportion to the size of the veins and points to the long-continued flow of solutions through the fissures.

The relative importance of cavity filling, replacement, and the force of crystallization as factors in forming the quartz veins is an unsolved problem. The positive evidence, as pointed out on page 45, favors the predominance of cavity filling. Nevertheless the gross effect of the appearance of certain veins as viewed in the stopes or crosscuts is that they are the results of replacement. This strong suggestion of replacement origin is well shown by the quartz ore of some of the stopes in the Argonaut mine, as illustrated in Plate 5, A, which suggests quartz containing innumerable residuals of unreplaced slate. But the evidence as seen under the microscope is firmly against the idea of replacement of the slate by quartz.

Becker[14] in 1905 suggested that the Mother Lode veins were opened by the force of crystallization of

the growing quartz. Although it is highly improbable that the feeble force of crystallization could have crowded apart the walls of the veins to make room for the growing quartz, it is easily conceivable that after breccias had been formed in the fault fissure, or belts of schistose slate, or open spaces the force of crystallization thus working against no great resistance might pry apart the leaves of the slate and produce such closely spaced ribboned ore as is shown in Plate 4, A. But no criteria are known by which it can be established that the force of crystallization has been at work.

On the whole the theory that the Mother Lode quartz veins were developed by successive enlargements appears to fit the facts best. A later movement along the fissure might crush the earlier-formed quartz, and this would account for the fact that some of the gold occurs in crushed quartz. It is an interesting question whether such crushed quartz might not recrystallize under the influence of the later solutions and thus obscure or obliterate some of the evidence of successive movements along the fissure.

The great vertical extent of the Mother Lode veins indicates that they were formed by hot ascending solutions, as is shown also by their mineralogic character. They belong to the mesothermal group, formed under conditions of intermediate temperature, and the sporadic occurrence of pyrrhotite, tourmaline, and magnetite show that they are transitional to the hypothermal group. Probably the ore-depositing agency was hot water that carried in solution gold, silver, lead, zinc, and other heavy metals, also potassium, sulphur, arsenic, and carbon dioxide. The effect of the carbon dioxide and the potassium vastly exceeded that of the other constituents. The carbon dioxide liberated immense quantities of silica from the wall rocks, and this silica was delivered to the vein channels, where it was in part precipitated as quartz. The potassium was utilized in effecting the extensive sericitization of the wall rocks.

A question of much practical as well as scientific interest is what determined the very different behavior of the gold in the ore-forming solutions at different localities. In places the gold migrated freely into the wall rocks, forming the gray ore and the mineralized amphibolite schist ore. In places pyritized greenstone is valuable ore; in others, similarly pyritized and hydrothermally altered greenstone is valueless. Higher temperature of ore-forming solutions at some places than at others, the higher temperature giving the gold greater mobility, suggests itself as an explanation; but this suggestion appears to be negatived by a comparison with the mode of occurrence of the gold in the stringer lode of the Alaska Juneau mine at Juneau, Alaska; the gold there was deposited at a higher temperature than in the Mother Lode belt, but it is nevertheless confined to the quartz in the veinlets that ramify through the slates. It is

[11] Quiring, H., Thermoaufstieg und Gangeinschieben: Zeitschr. prakt. Geologie, Jahrg. 32, p. 166, 1924.
[12] Tronquoy, R., Contribution à l'étude des gîtes d'étain: Soc. min. France Bull. vol. 35, pp. 456–465, 1912.
[13] Spurr, J. E., The ore magmas, vol. 1, pp. 76–85, 1923.
[14] Becker, G. F., and Day, A. L., The linear force of growing crystals: Washington Acad. Sci. Proc., vol. 7, p. 283, 1905.

this restriction of the gold to the quartz veinlets that makes possible a rough sorting of the ore and permits the successful working of an ore body that averages 63 cents to the ton.

We come now to the source of the ore-forming solutions. Long ago Richthofen,[15] in a brilliant paper that still repays reading, maintained that the gold deposits of the Sierra Nevada were genetically related to the intrusive granite and were produced by emanations that issued from the deeper-lying portion of the magma, which was still fluid when that part of the granite now exposed to view had already solidified.

Ransome [16] thought that "in all probability the waters which carried the Mother Lode ores in solution were originally meteoric waters, which after gathering up their mineral freight in the course of downward and lateral movement through the rocks were converged in the fissures as upward-moving mineral-bearing solutions." In the light of present ideas this explanation may be modified as follows. The carbon dioxide, sulphur, arsenic, gold, and certain other constituents were probably supplied by exhalations from a deep-seated consolidating magma, as was also a part of the heat. These substances and the heat were added to the meteoric circulation, which supplied the necessary motive power to lift the liquid solution to the earth's surface.[17]

Which magma or magmas supplied the gold and other constituents is unknown. The attempt to ascribe their origin to gabbro, peridotite, and albite aplite dikes [18] finds no support from the mode of occurrence of the gold along the Mother Lode belt; in fact, the belt is particularly poor in gold where these intrusions are most abundant. It appears to be a safe conclusion, however, that the ore-forming emanations were given off during the final stage of the epoch of plutonic intrusion at or near the end of Jurassic time. For wherever the age of the gold deposits in the Sierra Nevada can be even approximately dated it proves to be younger than the granodiorite intrusions, and the community of characters shown by the gold deposits over the whole region is so great that there can be little doubt that they are all of essentially the same age. Even the copper deposits, most of which are west of the Mother Lode belt, are of postgranodioritic origin. Reid's ascription of the hornblendite at Copperopolis as the ore bringer of the great copper deposits there [19] has been proved erroneous by the subsequent discovery in the mine of large bodies of 4 per cent copper ore in the granodiorite, which is younger than

the hornblendite. Lest the idea of a zonal arrangement of the gold and copper deposits of the Sierra Nevada be seriously entertained, it is well to point out that copper deposits—for example, in the Noonday mine, in Eldorado County, whose ore consists of massive pyrrhotite intergrown with chalcopyrite—occur within the Mother Lode belt itself. Obviously the occurrence of this pyrrhotitic copper ore, so unlike the geographically closely associated gold deposits, was not determined by so simple a principle as zonal arrangement around a center from which ore-forming solutions emanated. The true explanation is part of the larger problem of the Sierra Nevada considered as a metallogenic province, but this problem is one that awaits field study.

THE MINES

PLAN OF TREATMENT

The mines are described in the following pages in geographic order from north to south. As in general only the operating mines were examined, descriptions of the many idle mines necessarily do not appear, but the mines here described are numerous enough to furnish examples illustrating in a detailed way the diversity of ore bodies that occur in the Mother Lode system. In describing the mines the main emphasis is put on their geology.

Valuable information on the location, ownership, equipment, and operating details of the mines will be found in the reports of the California State Mining Bureau. Particularly useful in these respects are the reports by W. B. Tucker in the State mineralogist's reports for the biennial periods 1913–1914 [20] and 1915–1916; [21] these two together constitute a convenient register of the active and inactive mines of the Mother Lode belt.

An account of the methods of mining used in the Mother Lode belt is given by Arnot.[22]

GEORGIA SLIDE MINE

The Georgia Slide mine is 1½ miles north of Georgetown, Placer County. It has long been considered as marking the north end of the Mother Lode belt. It has been worked more or less continuously since 1852, at first by bank blasting and sluicing and later by hydraulicking. As a result of these operations an immense pit has been formed, about 1,000 feet long and 600 feet wide at the top. It is estimated that about 3,500,000 cubic yards has been removed and that 2,000,000 cubic yards of tailings have accumulated in the canyon below the mine. Obviously these tailings must contain a large fraction of the gold originally present, and in 1915 a light 10-stamp battery was operated on them to determine this gold content.

[15] Richthofen, F. von, Ueber das Alter der goldführenden Gänge und der von ihnen durchsetzten Gesteine: Deutsche geol. Gesell. Zeitschr., Band 21, pp. 723–740, 1869.

[16] Ransome, F. L., U. S. Geol. Survey Geol. Atlas, Mother Lode District folio (No. 63), p. 7, 1900.

[17] Day, A. L., and Allen, E. T., The temperature of hot springs: Jour. Geology, vol. 32, pp. 184–185, 1924.

[18] Lindgren, Waldemar, Mineral deposits, 2d ed., p. 575, 1919.

[19] Reid, J. A., Econ. Geology, vol. 2, p. 414, 1907.

[20] Tucker, W. B., Mines and mineral resources of Amador County, Calaveras County, Tuolumne County, 180 pp., California State Min. Bur., 1915.

[21] Mines and mineral resources of the county of Eldorado, 37 pp., California State Min. Bur., 1917.

[22] Arnot, S. L., Mining methods in the Mother Lode district of California: Am. Inst. Min. and Met. Eng. Trans., vol. 72, pp. 288–304, 1925.

Since hydraulic mining was stopped small parties of lessees have made wages by working the quartz stringers or "seams," as they are locally called, for pockets. Some tunnels and drifts have been driven in the search for these pockets. There are no authentic records of output, but the best surmise is that a total of $3,000,000 has been produced.

The country rock is chiefly a deeply rotted green chloritic siliceous schist, with some interbedded black slate, striking N. 5° W. and dipping 80° E. The "seams" are quartz stringers that range in thickness from a fraction of an inch to a foot or more; the average stringer is perhaps 1 or 2 inches thick. Locally the stringers swell into lenses as much as 4 feet thick—exceptionally 15 feet thick. Five belts of schist in which the stringers are more abundant than elsewhere are recognized. The easternmost of these belts is 35 feet thick. The stringers consist mainly of coarse white quartz, but marginally they contain abundant albite and calcite. Free gold, some of it beautifully crystallized, and pyrite are the metallic constituents. Much of the gold is firmly attached to the quartz, and consequently most of it must have been lost during hydraulicking. The veinlets are almost invariably oxidized, hence they are honeycombed and highly limonitic; the adjacent schist is soft, oxidized, and rotten. Only in depth does the calcite appear in the veinlets. The gold, according to mint returns, is reported to have a value of $18.50 to $18.80 an ounce. Its indicated fineness is therefore 0.900, or considerably higher than that of the deep-vein gold of the Mother Lode system. An analysis of a specimen of free gold, given on page 37, shows its fineness to be 0.912.

GUILDFORD MINE

The Guildford mine is on the south side of the canyon of South Fork of American River, 3 miles north of Placerville. The claim was located in the fifties, and the mine was formerly known as Poverty Point. In 1915 a mill of 15 light stamps was in operation. In 1919 this mill burnt down as a result of forest fires, and the mine has been idle since then. The mine is opened by a series of adits. The main adit or haulage way, which is at the level of the mill, is known as the 400-foot level, because it is about 400 feet below the crest of the hill. A tunnel 500 feet vertically below the main haulage way and 150 feet above the river has been started.

The ore occurs in shoots distributed through a zone of schists and slates that rest on a well-defined footwall. The footwall rock is a schistose greenstone, well exposed at the portal of the main adit. It is homogeneous for a width exceeding 100 feet; it contained originally hornblende or augite phenocrysts, and these have been sheared out into dark-green glossy patches. Near the footwall of the ore zone it has been profoundly transformed into a light-colored rock composed of ankerite, albite, and sericite.

The footwall of the ore zone, separating the greenstone from the overlying schists and slates, strikes north to N. 20° W. and dips 50°–70° E. The slates strike N. 20° W. and dip 80° E. It is clear everywhere throughout the mine that the ore zone dips less steeply than the slates, the average divergence being 20°, and it is probable that the ore zone cuts the rocks at an acute angle along the strike.

Two shoots of ore were being mined above the 300-foot level in 1915. One rested on the footwall, and the other, which appeared to be the principal ore body, was 30 feet above the footwall. The ore consisted of slate or schist traversed by a network of quartz veinlets, commonly containing notable quantities of albite. Pyrite was the only sulphide. The ore was reported to carry $10 a ton. The sulphide concentrate was said to carry from $35 to $180, depending on whether it was obtained from the ore shoot or not.

In 1925 it was reported that the Guildford mine had been reopened and equipped with a 10-stamp mill, with the expectation that milling would start in the near future.[23]

PLYMOUTH MINE

History.—The Plymouth mine is at Plymouth, Amador County. It has had a long and eventful history, from many angles one of the most illuminating furnished by the Mother Lode mines. In the eighties the Plymouth Consolidated Gold Mining Co. was one of the largest gold-producing companies in California. It was formed in 1883 by consolidating the Empire, Amador Pacific, and Plymouth companies. Between the date of its organization and January 1, 1888, it produced $3,804,499, from which were disbursed 55 monthly dividends, amounting to $2,200,000. Before the consolidation the mines had yielded $2,500,000.

Two mills, aggregating 160 stamps, were operated; they were run by water power. The concentrate averaged from 1.25 to 1.50 per cent of the ore crushed; its value ranged between $100 and $200 a ton.[24]

The average yield per ton was $6.18 in 1886 and $7.59 in 1887, when 97,000 tons was crushed.[25] The total cost, including all expenses (mining, milling, treatment of concentrate, office, taxes, and prospecting) was estimated to be $3.07 a ton in 1887.

If, however, we apply to the figures for the year 1887 the test that the real cost is obtained by subtracting dividends from yield, we get the anomalous result that the cost per ton was $2.64 instead of the estimated $3.07. The inevitable conclusion follows that the income was disbursed too rapidly in the form of monthly dividends, so that insufficient sums were reserved for contingencies such as fire or pinching of the

[23] Eng. and Min. Jour.-Press, vol. 120, p. 103, 1925.
[24] Small, G. W., Notes on the stamp mills and chlorination works of the Plymouth Consolidated Gold Mining Co., Amador County, Calif.: Am. Inst. Min. Eng. Trans., vol. 15, pp. 305–308, 1887.
[25] California State Min. Bur. Eighth Ann. Rept., p. 44, 1888.

ore shoot. Both these calamities soon overtook the mine.

In January, 1889, fire broke out in the mine, and in order to extinguish it the mine was allowed to fill with water. When the mine was unwatered a year later, it was found that the fire had done much damage by causing extensive caving.[26] Soon after that time the mine was abandoned, and it remained idle until 1911.

It would appear from the history just given that the Plymouth mine had produced before its reopening in 1911 a total of about $7,000,000. The ore body from which that output was mainly won is described as "an immense chimney of ribbon quartz from 30 to 50 feet wide and 350 to 450 feet long." [27] This shoot is the famous Empire shoot, variously stated to have yielded $6,500,000 and $8,000,000. The probabilities, as usual, favor the lower of these traditional figures. The width or thickness given is manifestly excessive, for on using even the minimum figure, 30 feet, we find that the computed yield per ton of ore in the shoot is far too small. The results of this analysis of available historical data are typical of what happens to many other Mother Lode "data" when critically scrutinized.

After lying idle for 20 years the mine was reopened in 1911, on the initiative of Albert Burch [28] and W. J. Loring.[29] With British capital a corporation known as the California Exploration Co. (Ltd.) was formed, which undertook the preliminary work. Unwatering of the mine began late in 1911. The Pacific shaft, a vertical shaft 1,600 feet deep, was unwatered and reconditioned. It was found that the vein had been cut on the 1,600-foot level by a cross-cut extending 140 feet from the shaft and that the vein had been drifted on for 160 feet without finding ore, but that in the bottom of a winze that had been sunk to a depth of 75 feet below the level the vein assayed $17 a ton through a thickness of 28 inches. The winze was then deepened to the 2,000-foot level, all in ore, and development work was continued until 110,000 tons of ore assaying $6.35 a ton had been indicated.

Early in 1914 the Plymouth Consolidated Mines Co. (Ltd.) was organized, with a capitalization of 240,000 shares of £1 each, and purchased the mine from the California Exploration Co. for £204,482. A 30-stamp mill was built and began operating July 30, 1914. The total expenditure in reopening and equipping the mine was $960,000. Some adjacent claims—the Plymouth Gold Quartz, on the east, and the New London, on the south—were purchased in order to obviate any danger of apex litigation.

Operations continued profitably until 1920. With great regularity 11,000 tons a month was mined and crushed. About 125 men were employed underground and 60 on the surface. After 1920 rising costs and diminished ore shoots brought on a succession of lean years. Operating costs increased from $2.90 a ton, the average for the first two years after the reopening, to $4.97 a ton in 1920, or including development charges to $5.99 a ton. The shaft had been deepened by the end of 1924 to the 4,000-foot level (3,600 feet vertical depth).

On March 1, 1925 the mine went into the hands of a receiver and was offered for sale in the open market. It was bought by the Argonaut Mining Co. "for less than $75,000."

Development.—The mine is opened by the Pacific shaft, which is vertical down to the 1,600-foot level, below which it becomes an incline sloping on the average 57° E. The bottom level in 1924 was the 4,000-foot level, at a vertical depth of 3,600 feet. As in most Mother Lode mines, the upper, worked-out levels are largely inaccessible. Drifts on the vein rarely remain open for more than four to six weeks, and the heavy timbers, 18 inches square, that are used to support the roof are completely crushed. The timbering is one of the major items in the cost of mining. The policy of driving the levels in the foot-wall of the vein (in places in the hanging wall) was adopted to some extent; such drifts stand without timbering. On the 1,200-foot level 400 feet south of the shaft a long hanging-wall crosscut was driven to cut the Reese & Woolford vein. A typical Mother Lode black-slate gouge 1 foot thick, with a fine hanging wall striking N. 25° W. and dipping 70° E., was intersected, but nothing of value was found.

The ore was treated in a 30-stamp mill, having a capacity of 350 tons a day. This mill departed notably from standard Mother Lode practice.[30] The weight of the stamps is 1,250 pounds, and their duty is from 12 to 15 tons a day, the ore being crushed to one-fourth inch. The pulp goes from the batteries to classifiers, from which the fines is sent to two Wilfley tables and the coarse product to two Hardinge mills. The tailings from the Wilfley tables are returned to the Hardinge mills. It was found necessary in general to use only from 20 to 25 of the 30 stamps. Maps, development data, and production records were carefully kept in unusual fullness of detail, and for the free use of this valuable material I am greatly indebted to Mr. W. J. Loring. Mr. O. H. Hershey and Dr. Malcolm Maclaren were the consulting geologists for the Plymouth Consolidated Gold Mines Co.

Output.—The total output to the end of 1924 is roughly $12,000,000. From August, 1914, to June 30,

[26] California State Min. Bur. Tenth Ann. Rept., p. 117, 1890.
[27] California State Min. Bur. Eighth Ann. Rept., p. 44, 1888.
[28] Min. and Sci. Press, vol. 121, p. 301, 1920.
[29] Reopening of the Plymouth mine and the results: Min. and Sci. Press, vol. 121, pp. 771-772, 1920.

[30] Caetani, Gelasio, The design of the Plymouth mill: Min. and Sci. Press, vol. 109, pp. 670-679, 1914.

1924, 1,020,845 tons of ore was milled. This yielded 13,827 tons of concentrate, having a gross value of $1,616,616, or $116.92 a ton, in gold. The total value of the gold obtained by amalgamation and in concentrate was $5,676,921, or $5.56 to the ton of ore milled. The bullion fineness is 0.800 gold and 0.190 silver; these are average figures, for when the gold decreases the silver increases. There is no evidence that the fineness increases in depth. Of the gold 70 per cent is obtained by amalgamation and 30 per cent in the concentrate. Since 1915 the amount of gold obtained in the concentrate has ranged from 24 to 32 per cent in different years, but it may fluctuate nearly as much in successive months.

The concentrate produced has averaged 1.35 per cent of the ore milled. By a change of milling practice in 1923 to increase the recovery of the sulphides, whereby a lower-grade concentrate was produced containing 40 per cent of insoluble matter, the concentrate increased to 1.92 per cent of the ore milled. This figure indicates, the small loss during concentration being neglected, that the ore as milled contains 1.13 per cent of sulphides. These sulphides from the deeper levels (3,650-foot to 4,000-foot) consist preponderantly of pyrite, with very minor amounts of galena, sphalerite, and chalcopyrite. Concentrate from the 1,600-foot to 2,300-foot levels contained a perceptible amount of arsenopyrite, but that the disappearance of arsenopyrite from the ore of the deeper levels is of any significance can not be affirmed.

Geology.—The main rock at the Plymouth mine is the black slate of the Mariposa formation. There is also some interbedded graywacke, or sheared sandstone, some conglomerate, and some greenstone, chiefly in the form of lenses of augite melaphyre tuff and breccia. The belt of augite melaphyre east of the mill is probably an intercalated lava flow; it contains numerous large prominent phenocrysts of augite in a roughly schistose grayish-green matrix. The general distribution of these rocks is shown on Plate 9. These rocks strike N. 20° W. and dip on the average 75° W., and as shown by the interbedded layers of graywacke and graywacke slate bedding and cleavage are parallel. In their westward dip they are unlike the rocks at most Mother Lode mines.

Some masses of diabase porphyry, probably dikes, occur in the southeastern part of the area mapped. They are albitized, as described on page 18. Their presence has no known significance in regard to the occurrence of ore.

The eastern part of the mapped area consists of glossy gray-green amphibolite schists, dipping 85° E. The contact between them and the Mariposa slate is a reverse fault, which dips 60° E. It was a zone of mineralization, as shown at the Chicago lode, in the southeastern part of the area, and at the Mayella lode, where there is at the contact a wide shattered zone containing much quartz and albitized rock.

Although black slate is the prevailing rock in the mine, yet in the deeper levels, north of the shaft, greenstone appears in the footwall of the Empire vein, the downward extension of the mass that crops out at the surface in the northwest corner of the mapped area. On the 2,900-foot level a footwall crosscut penetrates 260 feet into greenstone, chiefly varieties rich in augite phenocrysts; at the face is plainly a flow or intrusive carrying abundant porphyritic augite. The greenstone is overlain by a foot of conglomerate, which is interbedded with greenstone tuff and slate. On the 3,400-foot level, north of the winze 400 feet north of the shaft, greenstone forms the footwall of the vein; it is intensely hydrothermally altered to gray rock. Hydrothermally altered greenstone also occurs as the hanging wall of the Empire vein on the 2,150-foot level; it occurs where it would be expected if the belt on the surface were projected downward on a dip of 75° W., as shown in Figure 7. It consists of dolomite, quartz, albite, sericite, and pyrite.

There are three principal veins on the Plymouth ground—the Empire, Pacific, and Reese & Woolford. The Empire and Pacific are the productive veins. The Reese & Woolford vein has not been shown to contain ore, although so far as its appearance and favorable environment go it might contain valuable ore shoots. There is also, as already mentioned, a mineralized zone along the contact of the slates and amphibolite schists. The course of these veins, as determined by O. H. Hershey, is shown on Plate 9. The course of the Empire, by far the most productive vein, is marked for a distance of about 7,000 feet by the North, South, Indiana, New London, and Pioneer shafts. A cross section through the Pacific shaft is shown in Figure 7. The position of this section is shown by the line A–A' on Plate 9. The Empire vein cuts through the westward-dipping rocks at an average angle of 62° E.; the Pacific is considerably steeper—80° E. Notable, because so exceptional in Mother Lode mines, is the belt of conglomerate, 110 feet thick, that forms the hanging wall of the Empire vein for a distance below the 3,225-foot level; the extension of this conglomerate in the footwall of the vein is not shown in Figure 7, as the evidence was not obtainable.

Before the reopening of the mine in 1911 the main output came from the Empire shoot, which yielded $6,500,000. This shoot occurred in the Empire vein south of the junction with the Pacific vein; it cropped out 1,250 feet north of the Pacific shaft. Its stope length was 450 feet, and it pitched 45° S. The pitch length was 2,000 feet, and the shoot was remarkably regular down to a vertical depth of 1,500 feet. Since 1911 most of the development work has been done on the Empire vein, and the ore mined has come almost entirely from this vein, mainly from the downward extension of the Empire shoot, which, however, has lost the regularity it had above the 1,500-foot level.

The vein matter ranges from solid quartz to black slate stringered with numerous quartz veinlets. The quartz filling is in places markedly ribboned with slate filaments, but not all ribboned quartz is ore.

The vein is generally bounded by remarkably clean-cut, well-defined walls, which in places are almost mathematically perfect planes. These walls range in dip from 75° to 45°, although the vein as a whole dips 62° E. A thick gouge, consisting of ground-up black slate spotted with fragments of white quartz, may form either the hanging wall or the footwall, or gouge may constitute both walls, and in places strands of gouge may cut through the vein. The gouge swells

FIGURE 7.—Section through the Pacific shaft, Plymouth mine. 1, Mariposa black slate; 2, graywacke; 3, augite melaphyre; 4, amphibolite schist; 5, Empire vein; 6, Pacific vein; 7, Reese & Woolford vein; 8, conglomerate

and pinches and locally attains a thickness of at least 4 feet. Where the vein is inclosed on both footwall and hanging wall by gouge the quartz has been reduced to a sugary condition, thereby proving that it has been severely crushed by postmineral movement. The postmineral movement appears to have taken place mainly along the premineral fissure and has produced strong gouges, which have a predilection for wrapping around the thick lenses of quartz. In places stringer zones 10 feet or more thick lie above the hanging-wall gouge. Locally the walls of the gouge are wavy, corrugated, and deeply fluted. The longer axes of the flutings are parallel to the dip and thus show that the postmineral movement was mainly parallel to the dip. According to Maclaren, who has studied the mine in detail several

times, the gouge has not separated the lode nor doubled it on itself, so "its economic effect may be disregarded."

The shoots of ore that have been worked are shown on Figure 8. The downward extension of the Empire shoot is not as regular below the 1,500-foot level as it is supposed to have been from the surface to the 1,500-foot level. Instead of pitching south at 45° it has become vertical or even pitches steeply north. Details concerning this shoot have already been presented on pages 28–29.

The vein was lean or unprofitable from the 2,500-foot level to the 3,050-foot level. There was another blank of 400 feet from the 3,235-foot to the 3,650-foot level. The shoot found below the 3,650-foot level is of good grade—about $10 a ton—narrow and lenticular, and 180 feet long. The quartz is considerably crushed, is practically barren of sulphides, and contains sporadic slate inclusions surrounded by crusts of ankerite and albite. On the south end the shoot pinches down to a gouge-filled fissure. At the time of my visit it had been explored 50 feet below the 4,000-foot level. At this depth sheared conglomerate forms the hanging wall of the vein.

This bottom shoot, as it is called, was found by sinking a narrow winze from the 3,400-foot level at a point 400 feet north of the shaft. At this point a footwall vein joins the Empire vein, with the consequent occurrence of a large body of quartz carrying $5 a ton. For a time all development work and extraction of ore was done through this winze, and this accounts in part for the increase in mining costs. Finally, late in 1924, the shaft was extended to the 4,000-foot level.

It began to be felt at the mine that exploratory work had perhaps been too exclusively directed toward finding ore in depth and that more longitudinal exploration should be done. With this purpose the 1,400-foot level was extended north on the Empire vein.

According to Maclaren three shoots were worked in the Pacific vein, all south of the Pacific shaft. (See fig. 8.) They were followed up to the 835-foot level, where they frayed out upward into a stringer lode.

In the Plymouth mine ore shoots appear likely to occur at junctions; at lenticular expansions of the vein, the positions of which can not be predicted, as their cause is not known; and possibly at changes of formation, which appear to have been a factor determining the forming of a branch vein.

FREMONT MINE

The Fremont mine extends 4,200 feet along the Mother Lode belt and is 1½ miles north of Amador City. It is equipped with a 40-stamp mill. The mine was shut down in 1918 and except for a brief period in 1923 has remained idle. The property is developed by the Fremont shaft, extending 3,000 feet on an incline of 50°, and the Gover shaft, an incline 1,500 feet in length, which is 1,430 feet north of the Fremont; the workings are connected underground.

thick belt of augitic greenstone tuff and breccia. The greenstone belt, as shown underground, contains some intercalated lenses of slate. All rocks dip 80° E.

FIGURE 8.—Ore shoots of Plymouth mine

The Gover is the older; the Fremont shaft was begun in 1900.

The Fremont shaft is at the surface in Mariposa slate, 50 feet west of the contact of the slate with a

Two veins—a footwall vein, known at the mine as the Mother Lode vein, and a hanging-wall vein—determine the development and exploration policy. The Mother Lode vein is so termed because it contains shoots of quartz ore; the hanging-wall vein is known as the gray-ore fissure, because although itself not generally ore-bearing it is accompanied by bodies of gray ore where one or both of its walls are greenstone. But this distinction between the two veins is not vital, for gray ore is associated with the Mother Lode vein below the 1,350-foot level. The angle between the two veins, on the 1,100 and 1,350 foot levels, is 30°; the Mother Lode vein strikes N. 20° E. and dips 45° E., and the gray-ore fissure strikes N. 10° W. and dips 55° E.

The gray-ore bodies have yielded the bulk of the output. The largest body of gray ore was 200 to 300 feet long, was 40 feet wide from the 1,100-foot to the 1,350-foot level, and extended down to the 2,300-foot level. Its greatest width was reported to be 70 feet. This fine ore body was situated in the wedge end of a mass of auriferous pyritized greenstone lying between the Mother Lode vein and the gray-ore fissure at their intersection. The ore follows the gray-ore fissure, lying in the footwall of that vein, and between the ore

FIGURE 9.—Section through Gover shaft, Fremont mine

and the Mother Lode (footwall) vein is a body of coarse white quartz or quartz-cemented shatter breccia of greenstone splotched with silvery mica.

The gray-ore fissure is a well-defined quartz vein, generally narrow, containing a foot or so of gouge. Its formation was accompanied by much auxiliary fissuring and shattering. The dip of the fissure ranges from 40° to 65° E. and averages 30° less than that of the slate and greenstone that it cuts. In strike also it has an equally large range. In places—for example, 330 feet south of the shaft crosscut on the 2,100-foot level—the vein expands to large bodies of coarse white quartz, 6 feet or more thick, which have been stoped. The hanging wall here is slate, and the footwall is greenstone. At some places where such bodies of quartz occur small shoots of gray ore have been found in the adjacent greenstone. The gray-ore fissure on the 2,300-foot level strikes N. 20° W. and dips 45°–65° E. Near the south end of the level the fissure was filled with a ribboned quartz vein inclosed between a well-defined slate hanging wall and a greenstone foot-

wall. The quartz filling contains no sulphides, but the adjacent greenstone carries pyrite and arsenopyrite and formed a small body of gray ore. The main body of gray ore on this level was 100 feet long and 20 feet thick and was continuous with stope No. 2 on the 2,100-foot level.

The gold tenor of the gray-ore bodies is reported to be very uneven, ranging from a trace to $20 a ton. The average value of the ore mined in 1915 was not disclosed, but that mined in later years averaged $5 a ton. Arsenopyrite is said to indicate better-grade ore, but in general the value of the ore can be determined only by assay.

Although the distribution of the gray-ore bodies presents many enigmatic features two facts stand out clearly in the Fremont mine—(1) shattering of the greenstone favored the formation of gray ore, and the main body of this ore occurred at the acute intersection of two veins, where the conditions for thorough shattering were most favorable; (2) only the purer greenstones—that is, those free from argillaceous admixture—were altered to ore.

The geologic conditions at the Gover shaft are shown in Figure 9. The conditions in this shaft are unusually favorable for measuring the displacement along the fault fissure occupied by a Mother Lode vein; it amounts here to 375 feet. The shaft follows the vein practically all the distance down to the 1,500-foot level (1,073 feet vertical depth). To use the terms commonly employed along the Mother Lode belt, the Gover vein is a "slate" vein from the surface down to the 170-foot point, a "contact" vein from that point to the 600-foot level, and a "greenstone" vein below the 600-foot level. The slates, which normally dip 75° E., are remarkably flattened against the vein where the greenstone forms the hanging wall, being bent back so that they dip as low as 40° E.

The vein was narrow in the slate but widened in depth, 50 and 60 feet being commonly reached.[31] The ore, however, was of low grade, and that in the bottom levels did not pay to extract. No ore was being mined in 1915, and only some prospecting was being done.

The bottom or 1,500-foot level from the Gover shaft is connected with the 1,350-foot level from the Fremont, which was driven north along the gray-ore fissure. It was thereby shown that the great quartz bodies of the Gover were formed along the north extension of this fissure, or zone of fissuring, as it might more appropriately be called, for in places the hanging wall contains much coarse white quartz as far as 75 feet from the fissure.

TREASURE MINE

The Treasure mine is between the Fremont, on the north, and the Bunker Hill, on the south. It was under development in 1915, and subsequently a mill with a daily capacity of 150 tons was built. This

[31] California State Min. Bur. Eleventh Rept. State Mineralogist, p. 146, 1893.

plant is an innovation in Mother Lode practice, as no stamps are used, and the ore is reduced by crushers and Hardinge mills.[32] The mine was idle in 1924.

The developments consist of an incline 1,600 feet long, sloping 50° E., and a winze 870 feet long. The collar of the shaft is on the contact of greenstone and Mariposa slate. The greenstones on the ridge east of the mine are highly augitic varieties and are mainly coarse breccias, which form prominent outcrops. This belt of greenstone, as shown on Rancheria Creek just south of the Treasure shaft, is 1,750 feet wide; underground exposures show that it contains some intercalated slate.

A zone of quartz croppings extends N. 40° W. along the contact of the slate and greenstone from the present shaft to the old Treasure shaft, a distance of more than 800 feet. A quartz vein occurs in the greenstone belt on the summit of the ridge east of the mine. It trends N. 35° W. and dips 55° E. It consists of 2 to 4 feet of solid quartz inclosed between well-defined walls, but there is also much quartz in the adjacent greenstone.

The Treasure shaft is sunk on the vein at the contact, keeping just under the hanging wall, which is predominantly greenstone. The structural relations show that the fissure occupied by the vein is the result of a reverse fault of not less than 300 feet displacement. The vein, dipping eastward 20° less steeply than the greenstones, penetrates in depth far into the greenstone belt; on the 1,000-foot level (as measured on the slope distance) there is already 400 feet of massive greenstone in the footwall of the vein, as demonstrated by a long crosscut into the footwall country rock.

The vein as a rule is thin and is accompanied by a thick gouge spotted with crushed quartz. It is not the vein filling, however, which was counted on to supply the mill, but the shoots of gray ore that occur from place to place in the footwall of the vein. These did not exceed 5 feet in thickness. Down to the 2,000-foot level gray ore had been found in the hanging wall of the vein in only one place. The gray ore, as in the adjoining mines, is an auriferous ankeritized greenstone carrying several per cent of pyrite and minor arsenopyrite.

BUNKER HILL MINE

The Bunker Hill mine is half a mile north of Amador City. It extends for 2,587 feet on the lode and adjoins the Treasure mine on the north and the Original Amador mine on the south. It is equipped with a 40-stamp mill. The tailings from the mill, which ran $1 a ton, were flumed to a custom cyanide plant, which treated them for a royalty of 25 per cent. According to Mr. Elisha Hampton, for many years the able superintendent of the Bunker Hill mine, it produced $1,272,000 before 1891, when it had attained a depth

of 700 feet on the incline. From 1899, when the mine was reopened, to the end of 1914 it yielded $2,863,000. The total output to the present time is in round numbers $5,000,000. Dividends amounting to $1,000,000 were paid up to September, 1916, after which payments ceased. Since 1916 operations have been much curtailed, and in 1924 the mine was idle.

The mine is developed by a double-compartment incline 2,800 feet long, sunk on an angle averaging 58°. Most of the workings are north of the shaft. In 1920 a winze was sunk 640 feet from the 2,800-foot level at a point 1,040 feet north of the shaft, but the ore found averaged only $3 a ton.

The principal vein is the Bunker Hill vein, or, as it is also known, the hanging-wall vein, because other veins, including the "gouge vein" and the Last Chance vein, lie in its footwall. Although the Bunker Hill vein and to a less extent the gouge vein have supplied quartz ore, the main dependence in recent years has been on gray ore.

The Bunker Hill vein at the surface is a contact vein, lying between black slate, which forms its footwall, and greenstone, which makes its hanging wall. The greenstone consists of augitic tuffs and breccias containing some interlayered hard black slate. In depth, at about the 800-foot level, the vein leaves the contact and enters the greenstone. It cuts through the greenstone layers and intercalated slates at an angle of 25° on both the dip and the strike.

As seen on the 2,200-foot and 2,400-foot levels, the Bunker Hill vein ranged in thickness from 2 inches to 25 feet; for long distances its filling consisted of gouge, locally as much as 5 feet thick, or in places of great lenses of quartz attaining a maximum thickness of 25 feet. These thick quartz lenses carried only a dollar to the ton. In places the hanging-wall country rock is closely stringered with quartz veins to a distance of at least 35 feet from the vein.

Some mineralized black slate, heavily charged with pyrite and especially with arsenopyrite, occurred in the footwall of the Bunker Hill vein at the north end of the 2,200-foot level and unexpectedly proved to be ore; some large stopes were mined from it.

The gouge vein, so called because its quartz filling has been reduced to the condition of a spotted gouge, is, along the line of the shaft, in the footwall of the Bunker Hill vein; north of the shaft the gouge vein inclines toward and joins the Bunker Hill vein.

In the wedge-shaped mass between the two veins was a large body of gray ore. It was 100 feet long and 25 feet wide and extended at least 350 feet on the dip. At the south end of the ore body—that is, in the direction away from the junction of the two veins—the ore became quartzose and of low grade. The "gray ore" changes abruptly around the edges of the ore body to green rock. Grayness is a characteristic of the ore, but the value of the gray ore can not be

[32] Logan, C. A., Amador County: California State Min. Bur. Seventeenth Ann. Rept., pp. 411–412, 1921.

estimated by visual inspection, even by those most familiar with it.

The geologic features of this part of the mine are summarized in Figure 10, which is a section drawn at right angles to the Bunker Hill vein 800 feet north of the shaft, or 100 feet south of the junction of the Bunker Hill vein and the gouge vein on the 1,400-foot level. Crosscuts into the footwall and the hanging wall country rocks allowed the geologic features to be ascertained, and the section has been drawn so as to show the information thus obtained. The section shows plainly that below the 1,200-foot level it is the gouge vein that has become the contact vein and

FIGURE 10.—Section through the Bunker Hill mine 800 feet north of the shaft. 1, Slate; 2, greenstone containing interbedded slate; 3, greenstone; 4, gray ore

separates a wide belt of black slate from an overlying belt composed chiefly of greenstone though containing some interbedded slate. The gray ore occurs in the wedge end formed by the convergence of the two veins. The section plainly suggests that the gray ore tends to follow the gouge vein rather than the Bunker Hill vein and that consequently the gouge vein merits more attention than has been given to it.

Another gray-ore shoot was found at the north end of the 1,200-foot level. It averaged 10 feet thick, in places expanding to 30 feet, and extended within 50 feet of the 800-foot level. The position of this shoot was evidently not determined by a junction, as it was

considerably north of the junction of the gouge vein with the Bunker Hill vein.

The gray ore is ankeritized greenstone containing 3 to 4 per cent of pyrite and minor arsenopyrite. The ore as mined in 1914 ranged from $3.76 to $7.71 a ton. The $7.71 average was that of a block of 1,270 tons taken out from the 1,400 east gray-ore stope. The sulphide concentrate from the gray ore ranged in gold content from $85 to $90 a ton.

ORIGINAL AMADOR MINE

The Original Amador mine is in Amador City, on the north side of Amador Creek. It extends for 1,450 feet on the Mother Lode belt and lies between the Bunker Hill mine on the north and the Keystone mine on the south.

A rather full account of the early history and prospects of the mine is quoted by Raymond [33] from a report by J. D. Hague. In the early seventies it was owned by the London & California Co., but the financial results of the working of the mine during 1873 were keenly disappointing to the stockholders, who expected annual profits of £76,800. For this reason the mine was shut down.[34]

After a long idleness the mine was reopened by the Original Amador Consolidated Mines Co. in 1908, the shaft was deepened to the seventh level, and the crushing of ore was begun in 1909. The mine had been bought under an option for $35,000, which covered only the Original Amador claim, of a linear length of 1,450 feet. Subsequently adjacent claims—the East Amador on the east, the Last Chance on the west, and others—were acquired, being paid for out of the earnings of the mine. In 1915 the milling capacity was increased to 300 tons a day, and as 1,500,000 tons of ore was blocked out above the 700-foot level, a prosperous future was foreseen.[35] Shortly afterward, however, difficulties in disposing of the tailings, mounting costs, and the low grade of the ore forced a shut down, and the mine has been idle since 1920.

The mine is opened by an inclined shaft 1,238 feet long, with 9,719 feet of drifts, 7,023 feet of crosscuts, and 27,465 feet of raises. The 700-foot level, the bottom level at the time of my examination in 1915, is at a vertical depth of 589 feet. The shaft is sunk on the vein from the surface down to the 300-foot level, where it passes into the footwall country rock of the vein.

[33] Raymond, R. W., Statistics of mines and mining in the States and Territories west of the Rocky Mountains for 1872, pp. 40–43, 1873.

[34] Idem for 1873, p. 88, 1874.

[35] O'Brien, T. S., Amador Consolidated milling plant, Amador City, Calif.: Eng. and Min. Jour., vol. 100, pp. 255–257, 1915.

The collar of the shaft is a short distance west of the contact of the Mariposa slate and a thick belt of augitic greenstone breccias and tuffs. Some of the more striking geologic features of the mine are shown in Figure 11. The marked flattening of the vein as it passes from the belt of black slate into the greenstone is the most notable. The reverse faulting along the vein fissure is probably more clearly and accurately measurable in the Original Amador than in any other Mother Lode mine; it amounts to 180 feet.

The ore mined was obtained in part from the Original Amador quartz vein and in part from large irregular masses of auriferous greenstone occurring in the hanging wall of the quartz vein. The vein along the line of the shaft is represented by two branches, which are separated by a thick mass of greenstone. On the 500-foot level this mass is 120 feet wide at the shaft. The footwall branch of the vein, known as the west or footwall vein, is thinner and of better grade than the hanging-wall branch. In the upper levels it had been largely mined out by the early operators and averaged probably $5 a ton.[36] It swells and pinches abruptly, ranging from a tight fissure to 5 feet of solid quartz.

The hanging-wall branch, the east or hanging-wall vein, is much the thicker, in many places being 15 feet or more thick. On account of its low dip its width in such places exceeds 30 feet; the maximum width seen was 90 feet. It is markedly lenticular and within a short distance of so great a width may pinch down to a mere seam. The filling is a coarse white quartz, carrying few or no sulphides and in places containing a multitude of inclusions of slate and greenstone. The sparse sulphide content, chiefly pyrite, is localized in these inclusions. The quartz is rather commonly splotched with aggregates of silvery white mica, which have been formed by the hydrothermal alteration of inclosed filaments of schistose greenstone.

The hanging-wall and footwall veins unite 220 feet north of the shaft on the 500-foot level. On the next level below this the junction is considerably farther north, evidently owing to the much flatter dip of the

hanging-wall vein (40°) than that of the footwall vein (60°). This junction appears to have favorably influenced the formation of ore, for some stopes were taken out here on the footwall vein.

At the south end line on the 700-foot level, where it is continuous with the "contact" or Springhill vein of the Keystone mine, the vein is 8 feet thick, but only the top 2 or 3 feet was extracted. This portion of the vein contained numerous angular inclusions of slate, which had come from one of the layers of slate interstratified in the greenstone breccias and tuffs.

In places the hanging-wall greenstone is traversed by a network of quartz veinlets, is ankeritic and pyritic, and contains sufficient gold to constitute ore. Extensive irregular masses of such "gray ore" were

FIGURE 11.—Section through the main shaft of the Original Amador mine. 1, Black slate; 2, Original Amador vein (the shaft was sunk on the vein to the 300-foot level); 3, greenstone; 4, black slate, in part banded with green slate

mined and made up the bulk of the output in 1915. The average gold content appears to have been at most $3.50 a ton, as the tailings of the ore mined in 1914 are reported to have carried 35 cents a ton, the extraction having been 90 per cent.[37]

Well shown in the Original Amador mine but rare in other Mother Lode mines are the so-called cross veins. They were particularly well exposed on the 700-foot level. Their courses, although in detail irregular, were in general transverse to that of the hanging-wall vein. Their length ranged from 40 to 90 feet. Beginning at

[36] Raymond, R. W., op. cit. for 1872, p. 41, 1873.

[37] Tucker, W. B., Mines and mineral resources of Amador County, Calaveras County, Tuolumne County: California State Min. Bur. Fourteenth Rept. State Mineralogist, p. 40, 1915.

their junctions as bodies of solid quartz, 8 feet or more thick, they abruptly dwindled and at their distal ends were filled with country rock. They crossed contacts without causing any displacement.

KEYSTONE MINE

The Keystone mine, at Amador City, is one of the most famous of Mother Lode mines. It was first worked in 1852 and was operated almost continuously until its shutdown in 1920. As early as 1854 the Keystone was paying monthly dividends of $200 a share, and in March, 1855, it paid $550 a share.[38] Raymond,[39] writing in 1877, said that it had become the leading mine of the Mother Lode belt because of its great production, large-scale operation, and economi-

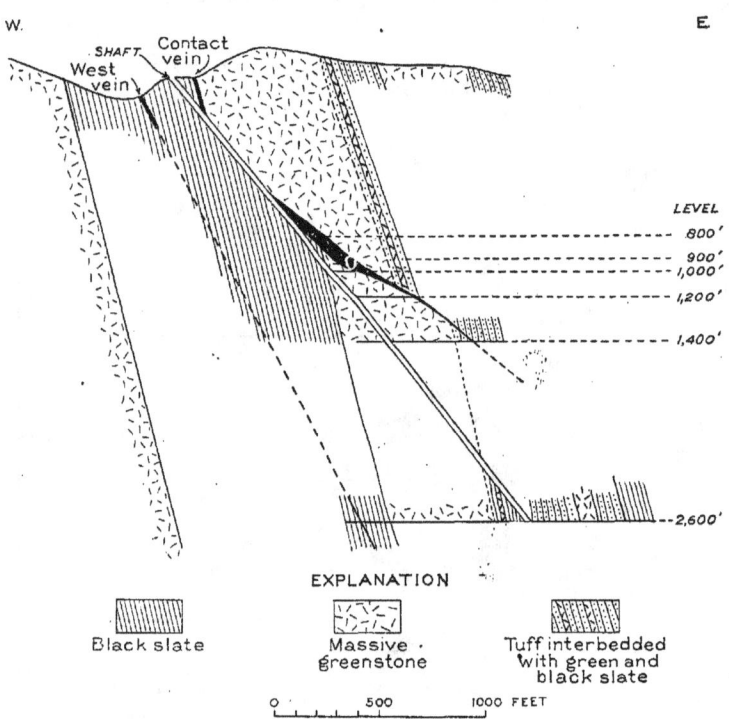

EXPLANATION

Black slate   Massive greenstone   Tuff interbedded with green and black slate

0   500   1000 FEET

FIGURE 12.—Section through the Patton shaft of the Keystone mine

cal management. Since 1870 its output had averaged $1,000 a day. The ore mined in those prosperous days yielded over yearly periods as much as $18 a ton.

According to Mr. R. C. Downs, for many years manager of the mine before it was closed in 1920, the authenticated output is $17,000,000. Of this amount $15,000,000 came from veins in slate and the remainder from the contact vein. Most of the ore mined in recent years has been obtained from the contact vein and averaged less than $3 a ton.

The mine is equipped with a 40-stamp mill. The duty per stamp during 1915 was 5.5 tons a day, and

the quantity crushed monthly was 6,200 tons. The following mill report is representative:

Report of mill of the Keystone mine, August 1–September 1, 1915

Ore crushed_____tons__	6, 502
Concentrate produced_____do___	268. 7
Average assay value of concentrate_____	$45. 73
Gold bullion value per ton of ore_____	0 56
Gold in concentrate per ton of ore_____	1. 89
Tailings loss_____	. 24
Total value per ton of ore_____	2. 69

The mine is opened by a three-compartment shaft—the Patton shaft—2,680 feet long and inclined at an average angle of 50°. The bottom level (2,600-foot) is at an approximate vertical depth of 2,000 feet. Between the 1,400-foot level and the 2,600-foot level no levels had been driven at the time of visit, and all ore was being hoisted from the 800, 900, 1,000, 1,200, and 1,400 foot levels. The ore was of extremely low grade, averaging $2.50 a ton, and the gold, unlike that in most Mother Lode mines, was largely contained in the concentrate. No profit was made in mining ore of that grade, but the operation of the mine without loss was a notable achievement.

When the shaft was deepened to the 2,600-foot level, it was hoped that the crosscut on that level would intersect a new ore shoot on the west vein, which had been very profitable on the upper levels. The supposed downward extension of that vein was indeed cut but was here only a gouge-filled fissure in the black slates. Insufficient funds did not permit the vein to be thoroughly explored, and operations were perfo ce centered on mining the known low-grade ore in the contact vein.

The Patton shaft is sunk between the contact or Springfield vein and the west vein. At the surface it is 90 feet west of the outcrop of the contact vein and 130 feet east of the outcrop of the west vein. The general geologic features of the mine are well shown by a vertical section through the shaft in Figure 12. The west vein crops out in a belt of Mariposa black slate, which, although lying between two belts of augitic greenstone breccias and tuffs, is itself free from volcanic admixture. This vein is reported by Mr. R. C. Downs to have yielded an authenticated output of $15,000,000. The west vein is supposed to have been intersected by the west crosscut from the bottom of the Patton shaft, as already related, and this supposition is shown in Figure 12 by the dashed

38 Trask, J. B., California State Geologist Fourth Ann. Rept., p. 60, 1856.
39 Raymond, R. W., Statistics of mines and mining in the States and Territories west of the Rocky Mountains for 1875, p. 47, 1877.

line joining the outcrop of the west vein with the vein cut on the 2,600-foot level. All workings, however, are inaccessible. The ore worked in early days was of high grade; the average yield during 1875, for example, was nearly $17 a ton. No ore was found in the slates below the 1,000-foot level. According to Ransome [40] the ore bodies in the slates were a series of parallel stringer leads, which dipped eastward into the main contact vein at angles of about 47°. "When carefully followed up, these leads are found to expand from a thin seam of gouge * * * into larger lenticular ore bodies, consisting of many large stringers with crushed and impregnated slate. The middle portion of these lenses is generally occupied by a large and solid quartz vein." This description renders somewhat doubtful the belief that the west vein has been cut on the bottom level.

The contact vein leaves the contact below the 800-foot level and passes wholly into the massive greenstone. It flattens greatly, dipping as low as 20° for great stretches. Enormous bodies of coarse white quartz occur here, some as much as 100 feet wide (equivalent to a thickness of 54 feet), but they carry less than $1 a ton. In spite of the immense bodies that occur in the fissure it becomes in places a mere seam, which is recognizable with difficulty as the main fissure. Thus for a long time the position of the vein in the long crosscut eastward from the shaft on the 1,400-foot level was doubtful. The crosscut was driven a total distance of 558 feet eastward, for it was not credible that the feeble fissure intersected at 440 feet was really the main "contact" vein.

Shoots of quartz ore were mined in the main vein. These occurred in the narrower parts of the vein; it was noticeable that in contrast to the barren parts the pay shoots carried considerable sulphide. It was believed that the vein contained more gold where it was more mixed with country rock; in fact, some of the ore mined was a breccia of angular fragments of slate and greenstone cemented together by quartz.

Three ore shoots were worked south of the shaft on the 800-foot level. The first one south of the shaft did not extend down to the 900-foot level. The second ore shoot, lying under a well-defined hanging wall dipping 40° E., was 25 feet or more thick and consisted predominantly of shattered greenstone ramified with quartz veinlets; there was no definite footwall, the gold content decreasing irregularly outward into the footwall country rock. The third shoot had a stope length of 240 feet. Another shoot of ore was being mined on this level at the north end line. The vein, about 5 feet thick here, was inclosed between well-defined walls dipping 45° E., and the quartz carried innumerable fragments of greenstone and black slate. This shoot is said to have extended with a slight northerly pitch from the surface almost to the 900-foot level.

"Gray ore," however, was the mainstay of the mine, and the term was used rather elastically to include all ore that was not quartz. The bulk of the gray ore, however, is ankeritized greenstone carrying disseminated pyrite and arsenopyrite. Microscopically the ore proves to consist almost wholly of ankerite with minor sericite and the sulphides. The abundant augite originally present in the greenstone has been transformed into ankerite of darker color than that in the surrounding matrix, so that the original texture of the greenstone is commonly fairly well preserved.

The gray ore occurs in bodies as much as 20 feet thick that form the hanging wall of the main vein. The ore bodies have assay walls, and as the rock becomes greener from the less amount of ankerite present it also carries less gold. Hydrothermal alteration of the greenstone, however, extends at least 100 feet perpendicularly from the vein. Shoots of gray ore of a stope length of 350 feet were being mined in 1915.

EUREKA MINE

The Eureka mine, at Sutter Creek, is one of the most famous of Mother Lode mines. It is variously known as Old Eureka (to distinguish it from the Central Eureka and South Eureka mines), Consolidated Amador, Hayward's, and Hetty Green. Its history has been recounted by Rickard.[41]

After having attained a depth of 2,063 feet on the incline it was shut down in 1886, having produced $12,000,000, according to the reasonable estimate by Rickard. It then lay idle for 30 years. In 1916 the property was purchased from Mrs. Hetty Green at a price a little over $500,000 in cash. The new company sunk the shaft to 3,500 feet (3,212 feet vertical) and drove levels at 1,700, 2,125, 2,950, and 3,500 feet. Crosscuts aggregating 1,500 feet were driven into the hanging and foot walls.[42] No ore was found, and in 1921 all work was stopped. In 1924 the property was bought by the Central Eureka Mining Co. in order to avoid threatened apex litigation, as the valuable ore shoots recently found in the deep levels of the Central Eureka pitched northward under its end lines and passed into the ground of the Eureka.

CENTRAL EUREKA MINE

The Central Eureka mine is 1 mile south of Sutter Creek, on the north edge of the broad upland between that town and Jackson. It lies between the South Eureka and the Eureka, and its original claim, the Summit, comprised only 1,160 feet along the Mother Lode. Because the end lines of the claim converge in

[40] Ransome, F. L., U. S. Geol. Survey Geol. Atlas, Mother Lode folio (No. 63), p. 8, 1900.

[41] Rickard, T. A., The reopening of old mines along the Mother Lode, California: Min. and Sci. Press. vol. 112, pp. 935-939, 1916.

[42] Logan, C. A., Amador County, California State Min. Bur. Seventeenth Rept. State Mineralogist, p. 411, 1921.

the direction of the dip, the length along the lode was becoming seriously shortened on the deep levels, but this adverse factor has now been removed by the purchase of the Eureka mine.

In 1895, when after long idleness the mine came under a new management, the shaft had been sunk 700 feet on the vein, but no ore had been found. On deepening the shaft ore was struck at the 1,100-foot

FIGURE 13.—Section through the shaft of the Central Eureka mine

level, which proved to be the top of a bonanza ore shoot. As the coming in of the ore coincided with the coming in of greenstone as the hanging-wall rock of the vein, this coincidence led to the formulation of the rule, emphatically maintained by some, that "it takes a hard hanging to make ore." But the later history of the mine has supplied the most brilliant refutation of the validity of that rule. Storms,[43]

43 Storms, W. H., The Mother Lode region of California: California State Min. Bur. Bull. 18, p. 64, 1900.

writing in 1900, regarded the mine "as one of great promise," and this promise was amply fulfilled during the next few years. Between 1901 and 1908 dividends amounting to $926,323 were paid.[44] A long period of lean years succeeded the prosperous epoch, and 48 assessments were levied. In 1915, at the time of my first examination, the contact vein, or hanging-wall vein, or east vein, as it was variously called, had become barren on the levels then being worked (3,000, 3,100, and 3,200), and the ore was being obtained from a series of so-called intermediate veins that occurred between the contact vein and the west vein and that were evidently the fillings of auxiliary fissures. During the next few years the mine was in peril of its existence, being at one time regarded as worked out. Fortunately, a new ore shoot was cut on the 3,500-foot level and has now been mined downward for 1,000 feet. From the proceeds of this shoot the hoisting plant has been modernized, expensive exploration work has been carried on, the adjoining Eureka property has been acquired for $150,000, and the payment of dividends has been resumed. The vein containing this ore shoot lies from 30 to 80 feet west of a well-defined but barren vein, known also as the contact vein, but this vein is not the contact vein of the earlier history of the mine.

The mine is developed by a three-compartment shaft inclined 70° E. The bottom level in 1924 was the 4,400-foot, which is 4,388 feet as measured on the incline, or 4,095 feet vertically below the collar. The ore is treated in a mill of 40 stamps of 1,280 pounds weight.

The output to the end of 1925 was about $6,000,000. During the year ending in April, 1925, 48,883 tons of ore was treated, from which was obtained $531,821 in bullion and concentrate, or $10.86 a ton.

The geologic section across the strike in the line through the shaft comprises, from west to east, (1) amygdaloidal augite basalt carrying prominent crystals of augite in a dark groundmass, apparently a fresh rock; (2) slate, 40 feet thick; (3) schistose greenstone tuff and breccia, 150 feet thick; (4) black slate, 400 feet thick; (5) augitic basalts and tuffs (greenstone). All these dip 80° E. At the surface the fifth belt is covered with a gravel of andesite boulders; it is, however, well shown in the hanging-wall crosscut on the 700-foot level, where it comprises 235 feet of massive greenstone—that is, augite melaphyre and tuff rich in fragmental augite—270 feet of alternating slate and greenstone, and then 60 feet of hard slates containing green laminae. At this point the supposed Railroad vein was cut, and what lies eastward is

44 Weed, W. H., Mines Handbook, vol. 15, p. 465, 1922.

unknown. The contact between the black slate belt and the overlying greenstone belt is heralded, as is common in the Mother Lode region, by the fact that at least 40 feet of the slate just below the greenstone is not a normal variety but is harder and laminated with green layers. Although on the 700-foot level there is at the base of the fifth belt 235 feet of greenstone without intercalated slate, on the 3,100-foot level the same greenstone is only 100 feet thick.

The shaft is sunk under the hanging wall of the hanging-wall vein as far as the 2,000-foot level, where it passes into the footwall country rock. As shown in Figure 13, the vein in the upper levels was inclosed in slate, but at a depth of 1,000 feet the hanging wall became greenstone, and at 1,100 feet the bonanza shoot came in, which extended down to the 1,900-foot level with northerly pitch. The maximum stope length of this shoot was 700 feet. According to Storms a considerable quantity of the ore averaged $70 a ton. Below the 1,900-foot level the hanging-wall vein became lean, and on the 3,100-foot level it was barren. It had flattened from an average dip of 70° E. to 50° or less on the 3,100-foot level, below which it has not been prospected. On the 2,800-foot and other levels down to the 3,350-foot level was cut the west vein, a gouge-filled fissure dipping 70° E. and carrying but a trace of gold. The black slates lying between the hanging-wall vein and the west vein are much disturbed and locally dip west, in places as low as 50°. The slates under the west vein, as shown by the 175-foot crosscut in its footwall on the 3,000-foot level, are quite regular. In the belt between the two veins occurred a number of the so-called intermediate veins or ore bodies. They supplied the bulk of the ore mined in 1915. The intermediate ore body on the 3,000-foot level, for example, was 330 feet long, ranged from 4 to 10 feet in thickness, and averaged $5 a ton. It stood nearly vertical, its walls were irregular, the adjacent slate was gashed with horizontal veinlets of vuggy quartz 1 to 2 inches thick, and as is characteristic of all such intermediate veins, there was no gouge. Another intermediate vein was stoped from the 3,200-foot level to the 3,100-foot level and higher; it averaged $3 a ton.

The ore shoot on which the present prosperity of the mine hinges was cut on the 3,500-foot level. The geologic relations exposed here have persisted with remarkable regularity to the greatest depth attained, as shown in Figure 14. The shoot persisted up to the 3,400-foot level, and the relation of the vein to the veins of the higher levels is lost in caved ground. At the mine the vein is regarded as the downward extension of the Railroad vein, which crops out in the Tanner ranch.

The only information on record about the Railroad vein is a single paragraph in "Mineral resources of the States and Territories west of the Rocky Mountains" for 1867, page 76, as follows:

The Railroad mine, 800 feet long, has been worked four years, has produced $70,000, and has had much rock which yielded $15 per ton. A depth of 340 feet has been reached, and drifts have been run 300 feet on the vein. There is no mill connected with the mine.

The vein containing the ore shoot that is now being worked, which may be called the ore vein, is 32 feet west of what is called the contact vein. On the deeper levels the interval between the two veins is as much as 70 feet, and the rock between them is highly disturbed black slate. The contact vein is a well-defined fissure, mostly filled with gouge and carrying in places $1 or $2 to the ton. Its hanging-wall rock

FIGURE 14.—Detailed geologic cross section of the lower levels of the Central Eureka mine

EXPLANATION

Black slate

Greenstone containing no intercalated slate

Greenstone alternating with slate

is greenstone alternating with slate, including tuffs or graywackes speckled with quartz particles.

The ore shoot where first cut on the 3,500-foot level was only 60 feet long but contained high-grade ore, in places assaying $50 a ton in 6-foot samples. All ore in 1924 was being hoisted from the 4,250-foot level, and the description of the ore shoot as seen there will suffice. The vein cuts the strike of the strata at an acute angle. The main ore body consisted of a lens of highly ribboned quartz, in which the ribbons were wavy or corrugated. It assayed about $20 a ton, though carrying little sulphide and that chiefly pyrite. The lens was 180 feet long, and its maximum width was 35 feet; it dipped 75° E. and pitched steeply northward. There were thick gouges on both walls, and the intervening quartz was much fractured and largely crushed to a sugary condition. The hanging-wall rock is black slate, greatly crumpled, and the footwall at the south end is black slate inclosing a 3-foot layer of greenstone; at the north end of the shoot the footwall rocks are banded black and green slates. At the south end of the ore shoot the quartz pinches out abruptly, but the gouge continues; in places the fissure is filled with 8 feet of shiny black gouge without quartz. The ore shoot lay north of the north end line of the Central Eureka; and it was the inherent difficulty that either claimant would have in establishing apex rights to the ore body that led to the compromise whereby the Central Eureka Mining Co. agreed to purchase the property of the adjoining Eureka mine for $150,000.

A detailed picture of the ore shoot as mined out above the 4,100-foot level is presented in Plate 10. The abrupt bulging of the ore shoot, which is one of its remarkable features, is shown, as well as the northward pitch of the shoot and the occurrence of the high-grade ore in the wide parts of the vein.

Some notable structural features are shown on the 4,250-foot level. The gouge of the contact vein merges with that of the ore vein 175 feet south of the shaft crosscut, thus showing that the two veins are in reality branches of one fissure. South of the junction the hanging wall rock is a greenstone schist, but the footwall for 120 feet south of the junction continues to be black slate, and then both walls become greenstone. Accordingly, the faulting that produced the main fissure has caused a horizontal displacement of 120 feet. Another notable feature shown is the transverse quartz vein in the footwall of the ore vein. (See fig. 14.) It trends S. 60° W. and dips 60° N.; its hanging wall is slate, highly corrugated and "graphitic," and the footwall is greenstone. Like all other greenstones adjacent to Mother Lode veins, it has been highly ankeritized and under the microscope is seen to have been originally a greenstone tuff. This transverse vein, which was drifted on for 250 feet, does not fault the ore vein and therefore appears to be most probably of contemporaneous origin.

Black slates without interlayered greenstones form the footwall country rocks on the upper levels, as uniformly shown by the crosscuts; in fact, on the 1,800-foot level a belt at least 300 feet thick has been crosscut. On the lower levels, however, greenstone is beginning to appear, for in the south footwall crosscut on the 3,700-foot level at least 200 feet of greenstone is shown.

SOUTH EUREKA MINE

The South Eureka mine is 1 mile south of Sutter Creek and lies between the Central Eureka on the north and the Oneida on the south. The vein system at this locality is covered with a thick mantle of coarse bouldery andesite gravel. According to Ransome

"the mine was located by means of surveys and through certain shrewd deductions drawn from the record of mines north and south of it. It is an example of bold yet legitimate and successful prospecting." The further history of the South Eureka shows that the records of neighboring mines should not be used exclusively as guides in exploration, because the vein system in each mine has its own idiosyncrasies. Moreover, it enforces the well-known principle that the belt in which a vein system occurs should be thoroughly explored by crosscutting. For, according to Mr. W. H. Schmal, superintendent of the South Eureka, operations during the first 18 years of the life of the mine were confined to the hanging-wall vein; but only about 1908, after the footwall vein had been discovered essentially by chance and contrary to expert advice, did the mine become prosperous and pay dividends.

In 1915 the mine employed underground 220 men; but it was obvious at that time that the known ore bodies were being rapidly exhausted. In 1917 the mine was shut down, but the pumping plant on the 1,800-foot level is used to drain the Central Eureka, and the shaft is maintained to serve as a second exit from the Central Eureka.

The mine was developed through an inclined shaft that is 2,730 feet long and attains a depth of 2,470 feet and by winzes extending down to the 3,100-foot level. At the 1,700-foot level the shaft strikes the hanging-wall vein, follows down in it to the 2,000-foot level, and then passes into the footwall country rock. Six men were continuously employed in repairing the shaft, mainly in keeping it in alinement.

The workings of the South Eureka are joined with those of the Oneida mine, which is owned by the South Eureka, the 2,100-foot level being connected with the 1,800-foot level of the Oneida. On the north the 2,700-foot level is connected with the 2,540-foot level of the Central Eureka.

A mill of 80 stamps, embodying in most representative form standard Mother Lode practice, was operated. The ore was crushed through a 24-mesh screen, and the pulp after passing over the amalgamation plates went without classification to 48 belt vanners. The extraction was 92 per cent.

During the fiscal year ended February 28, 1915, there was milled 99,892 tons of ore, which yielded $348,140 in bullion and $134,527 in concentrate. The ore yielded $4.83 a ton, and the tailing loss was $0.423 a ton, so that the content of the ore was $5.253 a ton. The concentrate, comprising 1.7 per cent of the ore milled, was composed chiefly of pyrite and quartz with minor arsenopyrite, galena, chalcopyrite, and zinc blende and carried $79.42 a ton. The operating cost per ton milled was $3.275; if all disbursements were included the operating cost was $3.577. These costs were regarded as abnormally high, because of reduced production and the expenses of new construction. The normal rate of production was 12,000 tons a month. But costs as low as these are not likely to be achieved again on the Mother Lode.

During 1915 the output was $582,764 from 145,124 tons of ore, or $4.34 a ton, and dividends of $125,354 were paid.[45]

The vein system is in a belt of black slates several hundred feet wide that lies between greenstones. The footwall greenstones are exposed at the surface 200 yards west of the shaft as massive rocks containing well-preserved stout prisms of augite. The hanging-wall belt is covered by gravel but in the deeper levels of the mine is seen to consist of pyroclastic greenstone alternating with slate. The principal veins are the hanging-wall vein and the footwall vein, so called before the fact became known that other veins occur in its footwall.

The black slate belt is 700 feet wide north of the shaft on the 2,000-foot level, its full width having been disclosed by a crosscut driven westward to cut the footwall greenstones; south of the shaft some greenstone occurs within the slate belt. The hanging-wall and footwall veins are roughly 250 feet apart on the lower levels, and the slates between them are highly disturbed, dipping west, contrary to the general eastward dip of all other rocks in the mine. The hanging-wall vein, dipping eastward less steeply than the inclosing rocks, enters the hanging-wall belt of greenstone just above the 2,000-foot level; on the 2,900-foot level there is already 70 feet of augitic greenstone under the vein. On entering the greenstone the vein flattens in dip from 70° E. to 50° E. Although a strong, well-defined vein, it carries no gold. A large amount of exploration that proved to be unprofitable was done on it, doubtless influenced by the fact that the bonanza ore shoot in the adjoining Central Eureka mine was in this vein.

The footwall vein was discovered by following one of the "intermediate" veins that extend diagonally across the belt between the hanging-wall and footwall veins back to its junction with what proved to be the productive vein of the mine; hence it became known as the main vein. Ore was obtained also from a number of intermediate veins. These veins have the same general character as those in the Central Eureka—that is, they have little or no gouge and are inclosed by rocks that are the same on both walls, thereby indicating that there has been no displacement along the fissures.

On the 1,800-foot level drifts were run on four separate veins. The westernmost vein, lying in the footwall of the footwall or main vein, was stoped continuously from the 2,000-foot level up to the 1,800-foot level. It was inclosed in greenstone on the 2,000-foot level, had a width of as much as 12 feet of solid quartz,

⁴⁴ Weed, W. H., Mines Handbook, vol. 13, pp. 572-573, 1918.

and averaged $4.60 a ton. As it entered the slate the tenor increased abruptly to $10 a ton, but the vein soon pinched and above the 1,800-foot level proved unprofitable.

The main ore shoot was only 60 feet long on the 2,900-foot level and averaged $7 a ton. It was near the Central Eureka end line. Ore was found to extend 140 feet below this level, but the vein then became barren.

On an option to purchase, the Central Eureka Mining Co. prospected the downward extension of this vein by workings driven from the company's 3,350, 3,900, and 4,100 foot levels, but it is reported that no ore was found. Obviously, however, as is enforced by the recent history of the Central Eureka mine itself and of the Old Eureka, the exhaustion of the South Eureka can not be regarded as proved.

KENNEDY MINE

The Kennedy mine, owned by the Kennedy Mining & Milling Co., is in Amador County, 1 mile north of Jackson. The claim was located in 1856 by Andrew Kennedy and was developed by a shaft near the north boundary line, which was sunk to find a continuation of the ore being mined in the adjoining Oneida claim, on the north. Work was started later on the vein now being mined, and ore was extracted down to the 1,200-foot level. This ore was of low grade, however, and the operators made but little profit. The present company obtained control in 1888 and after sinking 200 feet to the 1,400-foot level struck good-grade ore. The mine has been profitable since that time, except during the last few years, when fire necessitated flooding the workings and operations were curtailed. The years 1920 and 1921, also 1922 and 1923 to a less extent, were times of reduced production, but in 1924 conditions began getting back to normal.

A 100-stamp mill treats the average monthly output of 10,500 tons. The maximum quantity crushed was in 1912, when 172,200 tons was milled.

The mine is worked through a vertical shaft, 4,764 feet deep, from which a large number of levels have been driven. This shaft gives no trouble in maintaining its alinement, except where it passes through the vein. In July, 1924, mining was in progress on the three lower levels—the 3,900, 4,050, and 4,200 foot levels. The rest of the mine was inaccessible, except the long crosscuts on the 1,950 and 2,400 foot levels.

Notwithstanding the great depth, the maximum air temperature on the 3,900-foot level at the time of this survey was 86.5° F. On the 4,200-foot level, then recently driven, the average temperature was 1½° lower. The humidity, as in other deep Mother Lode mines, is nearly 100 per cent.

An average of 72,000 gallons of water was being pumped daily in July, 1924. According to the superintendent, Mr. James Spiers, this water is obtained mainly above the 500-foot level.

The geologic features of the Kennedy mine are, in general, simple. A quartz vein, which fills a reverse-fault fissure, cuts at an acute angle, both in strike and in dip, a series of more steeply dipping Mariposa slate and greenstones of Mariposa age. These inclosing rocks may be divided into six belts, three of which consist predominantly of slate and three of greenstone.

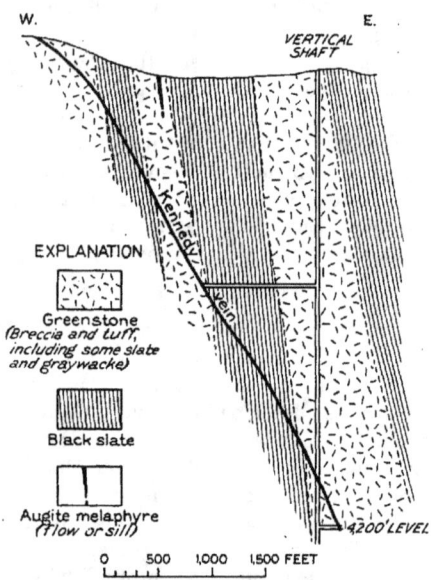

FIGURE 15.—Generalized section through the Kennedy mine along the line of the 1,950-foot crosscut

The vein crops out 120 feet west of the old North shaft on the Jackson-Martell road. As here exposed it is a thin and rather insignificant quartz vein, striking N. 10° E. and dipping steeply to the east. Black slate forms the hanging wall and mineralized greenstone the footwall. As the vein continues south, its dip flattens to 30° or less and greenstone forms both walls. The size of old open cuts on the vein testify to its greater width here than elsewhere.

The greenstones in which the vein crops out form the eastern border of a belt of augitic tuffs, breccias, and flows, which is 2 miles wide. The black-slate belt adjoining these greenstones on the east is from 275 to 300 feet thick and includes a few beds of gray-wacke. East of this belt is a second greenstone belt, 300 feet thick. This belt is made up of somewhat schistose tuffs and breccias, except for a highly augitic porphyry, which is not of pyroclastic origin but is a flow or sill. The porphyry is 44 feet thick and is 180 feet above the apparent base of the group. A second black-slate belt, 1,000 feet wide, includes numerous finely schistose tuffs. Two or more beds of fine-grained graywacke slate are exposed near its eastern boundary. A zone of transitional rocks separates these slates from a belt of 500 feet of schistose tuffs and breccias, within which are some black-slate layers. The Kennedy vertical shaft is in this belt.

These volcanic rocks are succeeded by 500 feet or so of rocks that were mapped by Ransome as Calaveras but are more probably black slates of Mariposa age, inclosing numerous tuff layers and beds composed in part of sedimentary material and in part of volcanic material. The wide silicified zone separating these rocks from the amphibolite belt on the east therefore represents a mineralized reverse fault that has thrust the amphibolite schists up over Mariposa slate and pyroclastic beds.

Underground the continuous and unweathered exposures of these rocks show that no sharp boundaries can be drawn between the belts that have been distinguished on the surface and named according to the predominant kinds of rock in them. Not only are the sedimentary and pyroclastic beds intimately interbedded in the three eastern belts, but sedimentary and pyroclastic materials are intermixed in individual beds, producing such intermediate varieties of rocks as green slates and augitic graywackes.

That the Kennedy vein in depth cuts through successive belts of rocks of differing character is shown in the vertical section of Figure 15. Because of the large

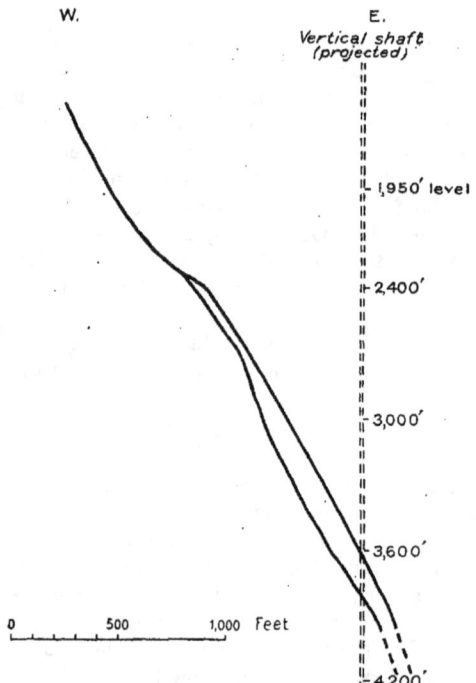

FIGURE 16.—Section through the Kennedy vein 280 feet north of the vertical shaft, showing the branching of the vein

quantity of slate and graywacke inclosed in the belt through which the vertical shaft penetrates, the boundaries between this belt and the adjoining belts are particularly unsatisfactory. The section is drawn from observations made on the surface, on the 1,950-foot level, and on the bottom levels and from information regarding the position of the vein shown on mine maps. The section shows the minor deviations in the dip of the vein and the notable flattening of the vein near the surface, where both walls are of massive greenstone. There is also a discordance between the strike of the vein, which averages N. 25° W., and that of the country rock, which is N. 50° W. The discordance is most apparent on the 3,900-foot level, where the vein, as it continues north, enters a belt of hard slates and bends nearly normal to their course.

A feature of the vein, not exhibited in Figure 15, is that below the 2,400-foot level in the northern part of the mine the vein splits into two branches. A partial section parallel to that of Figure 15 and 280 feet north illustrates this feature. (See fig. 16.) Examination of the junction of the two branches, known locally as the west and east splices, in the stopes proves conclusively that they are contemporaneous. The country rock in the wedge end is gashed by numerous quartz veinlets, and south of the junction the main vein is wider than normally.

The hanging wall on the three lower levels is sharp and smooth and is overlain by an uneven thickness of gouge. The gouge attains a maximum thickness of 6 feet, but in places it is entirely absent, as at the north end of the west splice on the 4,050-foot level. Angular pieces of the wall rocks are, in places, embedded in the triturated black slate that makes up the greater part of the gouge. According to the superintendent the gouge was on the footwall side of the vein in the upper part of the mine. It must, therefore, have crossed the vein at some point above the levels now being worked, as it is of postmineral origin. Water causes the freshly exposed gouge to swell, and this causes trouble in mining.

The footwall on the lower levels is not as well defined as the hanging wall, changing rather abruptly to a thin stringer zone. The black slates here commonly show a marked metasomatic alteration for distances as great as 10 feet from the vein. This effect of the ore-forming solutions is manifested by the occurrence of small pustules of ankerite between the cleavage planes.

The portions of the vein that are being mined are lenses, the largest 200 feet in length, and of diverse width, the maximum stope width having been 112 feet. The average thickness throughout the stope length is 15 feet. The lenses pass, within a very short distance, into a thin stringer lode, or even into a gouge vein without quartz filling. The stope plan of two lenses on the 3,900-foot level is shown in Figure 17. These lenses constitute the ore shoots and are remarkably persistent in depth, in view of their comparatively short horizontal extent. They pitch slightly to the south. In addition to the shoot on the main vein, both splices generally carry ore. The shoot on the east splice is the shorter. On two levels an alternation

of shoots occurred—that is, the shoot on one splice began where the other ended. All three shoots are present at or near the junction of the splices with the main vein.

Down the dip on these shoots is found a certain amount of zoning—that is, intervals of higher gold content alternating with those of lower. The upper 1,200 feet appears to have been the first zone of low-grade ore. This zone was followed by one of good ore that persisted for nearly 1,000 feet in depth. The succeeding poor zone was fortunately only a few hundred feet thick and was in turn succeeded by good-grade ore. The last few levels were again disappointing, but on the 4,200-foot level, the lowest level at the time of examination, it was thought that the top of another good zone had been reached, for the ore in this level is equal in grade to any previously mined.

The ore exposed on the three lower levels is chiefly a massive coarse quartz and carries a very small quantity of sulphides, which consist mainly of pyrite with insignificant quantities of sphalerite and galena and "traces" of arsenopyrite. Free gold is occasionally found. A silvery micaceous mineral, probably damourite, occurs locally in the quartz and usually indicates a good grade of ore. Galena was associated with some

FIGURE 17.—Stope plan of the Kennedy vein on the 3,900-foot level

specimen ore near the south end of the main vein on the 4,200-foot level. A fairly large proportion of the vein shows ribbon structure, particularly near the hanging wall. In places this ribboning is cut by veinlets of later quartz. Inclusions of the wall rock in the vein are locally abundant.

Of the 1923 output 76 per cent was obtained as bullion and 24 per cent from concentrate.

The concentrates produced during the last decade have averaged 1.25 per cent of the ore milled and carry $110 a ton. This percentage is perhaps higher than would be estimated from the exposures of quartz underground. Probably there is some dilution of the quartz ore by pyritized wall rock that is unavoidably taken out with the ore.

The gold bullion ranges in fineness from 789 to 837 parts per thousand, according to the bar tested. The first bar, as it is called, which is obtained from the battery amalgam, is of the highest fineness, whereas bars from the plates are lower. The figures for July 9, 1924, are, first bar 818 parts gold, 164 parts silver; second bar, 810 parts gold, 166 parts silver; third bar, 789 parts gold, 166 parts silver. Of course, each melting will show slight departures from these figures. An average value of the fineness of the first bar during 1924 was 830; in 1927 about 832. The fineness of the gold bullion in 1892, when the ore was obtained from the 1,600-foot level, ranged from 815 to 834. Consequently there has been no progressive change in the fineness of the gold with increasing depth.

The deeper development work that has been done since the preceding description was written has verified the optimistic forecast on page 65. The ore mined during 1926 on the 4,350-foot level improved notably in grade, averaging $10 a ton. Considerable visible free gold occurs in the ore, and sulphides are sparse. From 8,000 tons of ore 55 tons of concentrate averaging 80 per cent of sulphides was obtained, showing that the ore contains approximately 0.6 per cent of sulphides.

In August, 1927, I was able to revisit the mine and see the bottom levels. The shaft had been deepened to 4,764 feet, and preparations were being made to cut the station for the 4,650-foot level. The chief workings were then on the 4,500-foot level. The foot-wall branch of the vein was intersected 380 feet east of the shaft by the crosscut from the shaft, which cuts through rather massive greenstone. At the intersection the vein was 25 feet thick and consisted of coarse white quartz containing sericitic (damouritic) films parallel to the walls. The sparse sulphides were localized in these films. Only 6 feet of slate gashed with quartz stringers here separates the two branches of the vein. Under the footwall branch is a 60-foot belt of black slate cut by innumerable quartz stringers in places making up one-half the bulk. The bottom levels show, as did the higher levels, that the vein is ribboned only where it is narrow.

The ore shoot was 700 feet long, having lengthened northward as well as southward, and it had improved in quality. The ore as shown by samples taken during development work ran from $15 to $20 a ton. They are as good as any in the previous history of the mine and are highly encouraging as to the downward persistence of ore in the Amador section of the Mother Lode system.

Other changes have occurred. The hoist has been electrified, and the economies anticipated have been realized; the speed of hoisting, however, has been reduced to 1,500 feet a minute. The capacity of the mill has been reduced from 100 to 60 stamps, as it was found that when all inefficient members of the working forces are weeded out operations are more profitable, even if conducted on a reduced scale.

In 1927 payment of dividends was resumed, and all indications pointed auspiciously to a period of prosperity.

ARGONAUT MINE

The Argonaut mine, owned by the Argonaut Mining Co., is 1 mile north of Jackson and adjoins the Kennedy mine, which is its neighbor on the north.

The claim on which the mine is situated was patented as the Pioneer Quartz mine in 1871, but nothing more than prospecting of its surface was done before 1893. At that time a new company was organized and began sinking the inclined shaft now in use. This activity was stimulated by the discovery of good-grade ore that had then recently been made on the 1,400-foot level of the Kennedy mine.

By the end of 1915, when a depth of 3,300 feet had been reached, the output had amounted to 792,449 tons of ore, which had yielded $6,378,000, or an average of $8.50 to the ton. The dividends paid amounted to $1,800,000, or 28 per cent of the gross output, a ratio that compares favorably with that shown by any of the world's larger gold mines. The ore mined in 1915 yielded $765,509, or $13.02 a ton.[46]

In recent years two disastrous fires curtailed production for more than three years. In 1924 the recovered value of the ore was $9.01 a ton; in 1925, $7.38; and in 1926, $5.33.

The total output to the end of 1926 is estimated to be $14,400,000.

The mine is developed by an inclined shaft, which in 1927 had attained a vertical depth of 4,681 feet. From this shaft some 40 levels have been driven, but most of them, except those now used for mining or maintaining a second exit, are inaccessible. The deepest working in 1927 was the 5,400-foot level. A storage-battery locomotive was used to haul the ore on the 4,800-foot level.

The water pumped from the mine is 50,000 gallons daily. A 60-stamp mill treats the monthly output of 7,500 tons. It recovers 90 per cent of the gold by amalgamation and concentration on tables, and the tailings are sent to the tailings plant of the Amador Metals Reduction Co., where they are cyanided under contract. The sulphide concentrate amounts to 2 to 2½ per cent of the ore, indicating that the ore has a sulphide content of 1.6 to 2 per cent. The fineness of the bullion ranges from 0.835 to 0.840.

Like the Kennedy vein, of which it is the southern continuation, the Argonaut vein occupies the fissure of a reverse fault and in depth cuts through at an acute angle the same sequence of alternating belts of greenstone and slate.

The outcrop of the Argonaut vein is concealed throughout the greater part of the length of the claim by a capping of Tertiary gravel made up of andesite cobbles. Near the south end line, however, a large cut, 65 feet east of the powder magazine, shows much coarse white quartz, with numerous stringers that branch out from it horizontally or even dip westward. The country rock inclosing this quartz is a deeply weathered schistose greenstone. This quartz outcrop is presumably the top of the Argonaut vein, which here, as at the Kennedy, flattens notably near the surface as the result of refraction on entering the greenstone.

[46] U. S. Geol. Survey Mineral Resources, 1915, pt. 1, p. 224, 1917.

The same succession of slate and greenstone in the hanging-wall country rock as at the Kennedy mine occurs at the Argonaut. It is of interest to note that the flow or sill of massive augite melaphyre that occurs in the second greenstone belt east of the summit of the ridge appears to have been cut in the hanging-wall crosscut on the 1,690-foot level, indicating the possibility that this flow or sill can be used as a horizon marker.

In the section through the mine (fig. 18) the reverse character of the faulting along the fissure occupied by the vein is well shown. The black slate in the hanging wall of the vein on the 2,400-foot level can not be definitely accounted for, because the levels next above and below the 2,400-foot level are inaccessible. It may be either a lens in the greenstone, for such lenses are not uncommon, or it may have been brought into place by movement along the gouge vein that separates it from the greenstone on the east.

Comparison of the section through the Argonaut shaft with that through the Kennedy (fig. 14) brings out clearly the fact that the strike of the vein and that of the country rock do not coincide. In the area between the two sections the vein strikes more northeasterly than the slate and greenstone. The average strike of the vein at the apex has been determined to be N. 19° 14′ W. as a result of end-line litigation.

The displacement along the fault fissure occupied by the vein, as measured by the displacement of the slate-greenstone contact in the upper part of the mine, where the dip is relatively flat, is at least 120 feet.[47] On the lower levels where the dip is steeper, the throw is apparently much greater. This difference suggests that there is a strong lateral component in the total displacement.

The country rocks, as at the Kennedy, are composed of greenstone (augite melaphyre tuffs and breccias) and slate, with all the usual intermediate and gradational varieties between them. The finer-textured members have a well-defined foliation or cleavage, and in places the discordance between this foliation and the true bedding planes is as much as 15°.

The wall rocks are hydrothermally altered near the vein. The slates are least affected, but the development of knots of ankerite in them for distances of 5 or 6 feet from the vein is characteristic. Much of the greenstone has lost its original appearance almost completely owing to the wholesale development of ankerite. Where seen on the 2,400, 3,300, and 4,800 foot levels this ankeritized rock either forms one of the walls of the vein or is in close proximity to it. Remnants of the original pyroclastic structure may still be seen in much of the material. The rock is similar to the "gray ore" of other Mother Lode mines, and, indeed, it assays as much as $4 a ton here. Although locally called "schist" and considered to be

47 U. S. Geol. Survey Geol. Atlas, Mother Lode District folio (No. 63), p. 8, 1900.

distinct from the greenstone, it is merely a hydrothermally altered facies of that rock. Figure 19 is a geologic map of the 3,300-foot level and shows a body of this "gray ore" in the footwall of the vein. In a footwall crosscut from the 5,100-foot level the pyritized greenstone averages $1.05 through a thickness of 40 feet.

Two shoots of ore are now being mined—a north shoot, averaging 250 feet in length, and a south shoot, which, on the 4,650-foot level, is 160 feet long and in places more than 30 feet wide. The south shoot has a well-defined hanging wall of black slate, along which there is a gouge with its characteristic highly polished

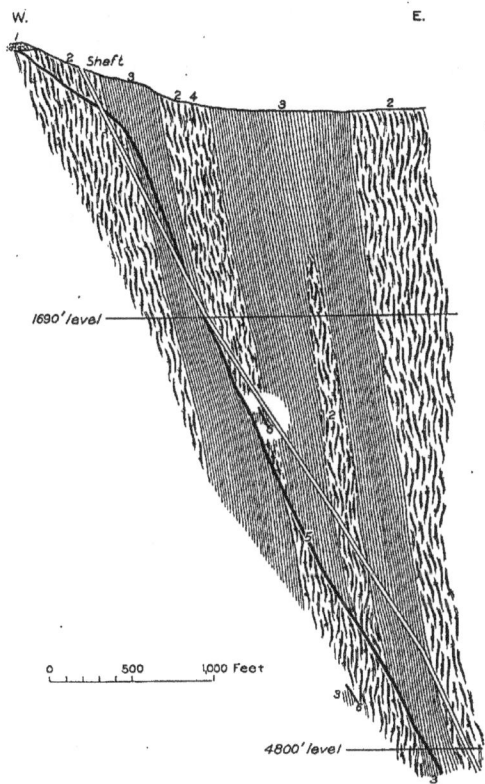

FIGURE 18.—Section through the Argonaut shaft. 1, Tertiary gravel; 2, greenstone; 3, black slate; 4, augite melaphyre (flow or sill); 5, Argonaut vein; 6, gouge veins

black surfaces. The slates above it are corrugated and crumpled. The footwall is indefinite, as the vein is underlain by a wide zone of stringers, which was said to assay $8 a ton over a width of 24 feet. The shoot bulges from a few feet to its maximum width within a short distance. This south shoot was a comparatively recent discovery. The north shoot, as seen just below the 4,200-foot level, differs from the south shoot in that it has a thick gouge on both foot and hanging walls.

From the surface down to the 2,530-foot level there was but one shoot, which lay to the north of the shaft, and it had practically no rake. This shoot has a maximum length of 600 feet and an average length of 300 feet. It was a continuation of the main Kennedy shoot and had essentially the same zones of lower-grade ore as were found in that mine. Below the 2,530-foot level the shoot raked 30° S. and terminated at the south end of the "gray ore" body shown in Figure 19. Below the 3,300-foot level a new shoot came in some distance north of this point and directly under (on the dip) the upper part of the old shoot. This is the north shoot that was being mined in 1924. The ore on these lower levels, particularly that of the south shoot, was of as good grade as any previously extracted.

The quartz in the south shoot was coarse, massive, homogeneous, and without noticeable sulphides. In appearance it resembled numerous barren veins, yet 5-foot samples assayed as high as $80 a ton. Free

Mokelumne Hill. It was first operated by a Captain Moore, but it had been idle for nearly 35 years when the present company began operations in 1921.

The mine is developed by a shaft inclined at 52°, which in September, 1924, was 1,190 feet long. Sinking was still in progress at that time. Mining or development work was being done on the four lower levels, the 640, 750, 800, and 950 foot. The ore is treated in a 20-stamp mill, built in 1922 at a cost of $51,000. The gold is won by amalgamation, 82 per cent being saved thus, and the remainder is obtained from concentrates whose tenor is approximately $100 a ton—about 5 ounces of gold and 2 ounces of silver to the ton.

Amphibolite and chlorite schists and members of the Calaveras formation are exposed in the vicinity

EXPLANATION

Black slate	Alternating slate and greenstone	Argonaut vein	Ankeritized greenstone or schist

0   50   100   200 FEET

FIGURE 19.—Geologic map of the 3,300-foot level of the Argonaut mine

gold is not uncommon, and a gray mineral thought to be petzite but shown to be galena occurs here. The ore of the north shoot is like that of other Mother Lode mines, being broadly banded by inclusions of black slate that are parallel to the hanging wall. Like the south shoot it also has a notably small content of sulphides.

On the 4,800-foot level the ore shoot was 950 feet long and attained a maximum width of 60 feet. The ore was of good grade, but below this level the shoot shortened somewhat and the grade of the ore diminished so that in 1927 the mine manager was confronted with the problem of getting through the poor zone that had been encountered.

MOORE MINE

The Moore mine, owned by the Moore Mining Co., is 1 mile southeast of Jackson on the county road to

of the mine. Green schists form the hanging-wall country rock of the vein at the surface, a fissile chloritic type forming the immediate wall. Black slate and quartzite of the Calaveras formation compose the footwall. Quartzite in what are apparently two beds is exposed west of the shaft, but when these beds are traced northwestward they are found to converge and meet at a point roughly 200 feet northwest of the shaft, which suggests that they are the outcrops of a single layer that has been closely folded. The evidence, because of poor outcrops, is not as full as is desirable.

The original cropping of the vein is now covered by the mine dump. There is a line of strong croppings of quartz, striking N. 40° W., several hundred feet northwest of the shaft, but it is unknown whether or not this quartz is the extension of the vein that is being mined.

The vein dips less steeply than its inclosing wall rocks. Down to a point somewhere between the 160 and 340 foot levels green schist continues to be the hanging wall and black slate the footwall. Below this point, however, both walls are the green schist. The vein therefore occupies the fissure of a reverse fault. The fault is well exposed in the north face of the drift on the 640-foot level, as at this point the vein filling is absent. The normal undisturbed contact between the green schist and Calaveras slate is exposed in the footwall on both the 340 and 640 foot levels and is seen to be transitional, as shown by the alternation of green and black layers. The dip is 65°–70° E.

The amount of the displacement along the fault can not be determined, as the slate-schist contact is not exposed in the hanging-wall country rocks. A minimum displacement of 200 feet, however, is required by the data available. The reverse character of the fault whose fissure is occupied by the vein is clearly brought out by a section through the mine. (See fig. 20.) A section along a line north or south of the one in Figure 20 would show the schist-Calaveras contact at a higher or lower point, respectively, as the vein cuts the bedding and planes of schistosity not only on the dip but on the strike.

The strike of the vein makes a small angle with the strike of the schists. This is well shown on the 640-foot level, where the vein, impinging on massive amphibolite schist, changes its course abruptly from N. 52° W. to N. 60° W. or more. The steepening of the dip from 52° E. to 70° E. below this level may also be due to the refractive effect of this same bed.

The ore body being worked in the summer of 1924 is a lens, fraying out at both ends along its strike and pinching vertically a short distance below the 800-foot level. Its horizontal length ranges from 100 to 200 feet, and the maximum thickness is 12 feet. The vein has generally a well-defined footwall, but on the hanging wall it shades into a stringer zone. The ore body rakes northwest at a steep angle.

The ore consists of coarse white quartz, carrying sparse pyrite and locally arsenopyrite and chalcopyrite. The concentrate obtained constitutes 0.8 per cent of the ore milled. Visible free gold is found in a few places. The veinlets branching off from the main body of quartz and cutting the adjacent country rock have albite bordering their walls, the smaller veinlets containing a relatively greater amount. The ore runs $9 to $10 a ton. The gold bullion has an average fineness of 0.819 gold and 0.171 silver.

Late in 1924 another body of ore was struck at 1,080 feet in the shaft. This ore body, dipping 70° E., consists of quartz with rather numerous inclusions of schist, some as much as 5 feet in length. Pyrite is distributed irregularly throughout, and a few specks of free gold may be seen. At a depth of 1,190 feet a thickness of 17 feet of ore averaging $14 a ton was reported by the superintendent, Mr. A. C. Wilson. The meager evidence available suggests that this new ore body is a lens similar in its general character to the ore body mined on the upper levels but situated somewhat farther in the hanging wall of the zone of fissuring produced by the reverse fault.

MAMMOTH MINE

The Mammoth mine, at the mouth of Spanish Gulch, on the north side of Mokelumne River, is typical of the pocket mines of the "black-metal belt." This "belt" lies half a mile or so west of the Mother Lode system and is so called because the gold is black when the ore

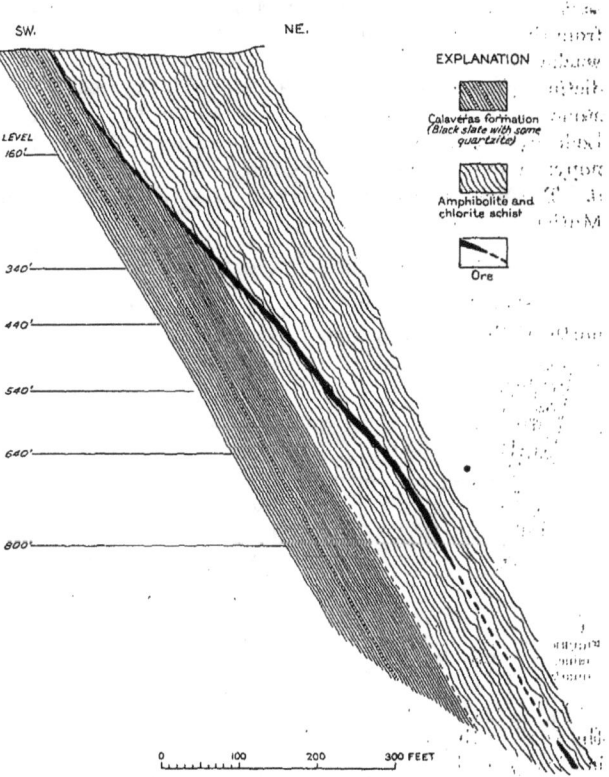

EXPLANATION

Calaveras formation
(Black slate with some
quartzite)

Amphibolite and
chlorite schist

Ore

FIGURE 20.—Section through the Moore mine on a line parallel to and 60 feet south of the 640-foot west crosscut

is pounded up in a mortar. The mine was being explored for pockets of ore in 1924 by the Federal Development Co., a force of 13 men being employed. Entrance to the mine is effected by means of a tunnel 3,000 feet long, trending N. 35° W.

The total output of the Mammoth mine is said to be between $300,000 and $400,000.

The country rock consists of black Mariposa slate and lenses of greenstone composed of tuff and breccia. Locally some lenses of coarse quartzose graywacke-slate occur. All these rocks strike N. 20° W. and dip 80° E. The ore occurs in a remarkably definite way.

It occurs in horizontal veinlets of quartz that are restricted to the vicinity of the contact of black slate and greenstone. (See fig. 21.) The veinlets follow horizontal joint surfaces, and most of them are but a few inches thick. Some are a foot thick at the contact, but invariably they diminish abruptly in thickness on entering the slate and pinch out. In the greenstone, however, the veinlets extend as much as 5 feet or even more from the contact. The ore consists of massive arsenopyrite in quartz; it pans free gold and may run as high as a dollar a pound. The ore is restricted to the quartz in the greenstone and as a rule extends not more than 2 or 3 feet from the contact; rarely as far as 4 or 5 feet. The veinlets may extend farther from the contact, but the arsenopyrite in them at greater distances from the contact is worthless. The occurrence of graywacke at the contact instead of black slate is said to be distinctly inimical to the occurrence of ore. The greenstone inclosing the ore shows the common Mother Lode hydrothermal alteration, but metasomatic arsenopyrite is the common sulphide disseminated through it. The smaller quartz veinlets, as is the rule in the Mother Lode belt, contain much albite.

AMADOR QUEEN NO. 2 MINE

The Amador Queen No. 2 mine is about 1 mile north of the Mammoth mine. It is developed by a

crosscut adit, which intersects a contact along which the pockets occur. Four small parties of lessees were working the mine in 1924. A large amount of work has been done by drifting on the contact both north and south of the adit and by stoping upon many of the veinlets.

FIGURE 21.—Pocket veins at the Mammoth mine. 1, Mariposa slate; 2, greenstone; 3, quartz.

Geologic conditions are essentially similar to those in the Mammoth mine in that the occurrence of the ore is clearly tied to a slate-greenstone contact. The quartz veinlets that contain the pockets are much more persistent, however. They dip about 45° and stand normal to the contact or roughly so. Many of them have been stoped from the tunnel level to the surface—distances ranging from 300 to 500 feet. Coarse gold in more or less well crystallized masses is contained in the arsenopyrite. Pieces weighing as much as half an ounce were shown to me and were said to be worth about $18 an ounce.

FORD MINE

The Ford mine is half a mile east of San Andreas. The property comprises 102 acres and extends 3,000 feet along the Mother Lode belt. The first mining was done in the nineties by the Ford Brothers, who

sunk a vertical shaft, the south shaft, 120 feet deep, on a large outcrop of quartz nearly 100 feet wide. Later the Ford Gold Mining Co. sunk an incline of 60°, the main shaft, at a point 300 feet north of the south shaft. It was deepened to 750 feet, and levels were turned off at the 100, 200, 300, 400, and 700 foot stations. A small shoot of high-grade telluride-bearing free-gold ore was cut on the 100-foot level. A 10-stamp mill was erected, but no vestige of it remains to-day. In 1900 the company became financially embarrassed, and the mine was shut down. It remained idle until October, 1922, when control was obtained by the Apex Mining Co. Since then moderate development work has been in progress on the 100 and 300 foot levels.

The prevailing rocks of the knoll on whose summit is the main shaft belong to the formation called amphibolite schist. Some 500 feet west of the shaft is a large mass of antigorite serpentine, which is intrusive into the Calaveras formation. The serpentine has been irregularly converted along its margin into a well-foliated talc schist, and the belt of talc schist at the old mill site, which is at least 110 feet wide, is probably an offshoot from the main body of serpentine and is continuous with it on the surface.

The most common rocks at the surface are gray-green schists streaked or mottled with white; under the microscope they are seen to be composed of albite, epidote, chlorite, sericite or talc, and amphibole, named in the order of abundance. They are feldspathic amphibolites, some of which contain only a minor amount of amphibole. Intercalated with them, 100 feet west of the main shaft, is a white schist containing small broken garnets. Two dikes of talc schist, called "talc dikes," cut these rocks. The contact of the amphibolite schists with the Calaveras black slates is 275 feet east of the main shaft.

Underground the geology appears more complex. The closely folded condition of the amphibolitic schists and associated rocks is well shown—in fact, better than anywhere else along the Mother Lode belt. As a result of the folding the strikes have a wide range, from N. 30° E. to N. 45° W. In the west crosscut on the 100-foot level a syncline of limestone beds plunging steeply northward is clearly seen; and at the face of the north drift from the east crosscut is revealed a fold in the amphibolite schist plunging 65° N., its axial plane dipping 70° E. Two talc dikes, one of which is 15 feet and the other a few feet thick, occur on this level near the east vein. A hornblendite dike was cut in the west crosscut of the 300-foot level. It is obviously somewhat sheared and contains discontinuous streaks of talc schist that were developed along zones of stronger deformation. It is doubtless good evidence on the origin of the so-called "talc dikes."

The closely folded condition of the rocks with their steep northward plunge causes the underground distribution of the rocks to be very different, even on the 100-foot level, from what would be inferred from their distribution on the surface.

In addition to the chloritic and amphibolitic schists, limestones, black slates, talc schist, and hornblendite that are seen underground, some light-colored rocks of obscure origin occur, as at the shaft station of the 300-foot level. They are probably related to the garnetiferous white schist already mentioned, but have been rendered more obscure by hydrothermal alteration, which has caused dolomite, sericite, and albite to form in them.

Four veins or lodes have been under exploration—the west vein, cropping out 120 feet west of the main shaft; the main vein, on which the main shaft was sunk; the hanging-wall vein, on which the south shaft was started; and the east vein, cropping out 270 feet east of the main shaft. The east and west veins are richer than the others and have therefore been the main objectives in recent years.

The west vein, as exposed at the surface, is a stringer lode 4 feet thick standing vertical. Its west wall is a sericitized amphibolite schist, and its east wall is talc schist. It has been cut on the 100-foot level, where it shows some free-gold ore.

As previously mentioned, a shoot of telluride-bearing ore was cut in the main vein on the 100-foot level. The tellurides found are petzite and hessite—two minerals between which there is no sharp boundary. As seen in the west crosscut of the 300-foot level, the main vein consists of a belt 75 feet wide, which is divided into two parts by a horse of schist 30 feet wide. It has a footwall of black slate, above which is a stringer zone of 30 feet, succeeded by 30 feet of dolomitized amphibolite schist, and then by another stringer zone 25 feet wide.

The hanging-wall vein has a wide quartz outcrop at the south shaft, but it can not be traced far on the surface. The lode, as seen in the east crosscut of the 100-foot level, is a belt of quartz stringers and albitized country rock about 20 feet wide. Much of the country rock has been converted into a white fine-grained aggregate of albite carrying 3 to 4 per cent of pyrite. An 8-foot cut across part of the belt is reported to have assayed $16 a ton. The lode has been cut also on the 300-foot level but is thinner there.

The east vein is a narrow quartz vein inclosed in black slate, which has been explored for a distance of 75 feet on the 100-foot level. It shows free gold in places. It intersects the two talc dikes previously mentioned and is reported to be particularly rich at the two intersections.

UNION MINE

The Union mine is 3 miles southeast of San Andreas. It has been idle since 1888, and none of the vein is now visible at the collar of the shaft, which was sunk on the vein. The mine is described here, however,

because it represents a large number of mines along the Mother Lode belt in the long time that they have been idle and in the difficulty or impossibility of judging their future prospects from existing surface showings, or from the available information on their history—information that could be proved or disproved only by considerable expensive exploratory work.

Early operations, when the mine was owned by the Rathgeb Brothers, are described in Raymond's reports.[48] The ore yielded $15 a ton, contained no sulphurets, and came from a quartz lode that averaged 5 feet in thickness. The depth attained was 150 feet. A 10-stamp mill was operated. The fineness of the bullion produced was 0.929 gold and 0.061 silver, a purity that indicates that the gold had been considerably refined in the zone of oxidation above water level. Current report is that the Rathgebs took out $175,000 from above the 120-foot level, which was about the water level. In 1886 they sold the mine to the Union Gold Co. (Ltd.) for £40,000, but three years later the mine reverted to them.

The vein, probably more accurately termed the lode, is said to be 30 feet wide. The hanging-wall rock is amphibolite schist; the footwall rock is a more slaty facies of the amphibolite schist formation. The mine was developed by an incline 455 feet long. It is reported that the work of the Union Gold Co. was almost wholly done in the hanging wall of the quartz vein instead of in the rich footwall schist zone. The ore extracted milled under $4 a ton and proved unprofitable.

GOLD CLIFF MINE

The Gold Cliff mine, owned by the Utica Mining Co., is in the west part of Angels Camp. It has been idle since 1920, and the following notes are based mainly on a brief visit in September, 1914. At that time a 40-stamp mill was being operated at a small monthly profit; 200 tons of ore was crushed daily, which yielded from 4.5 to 5 tons of concentrate containing $40 a ton in gold. The ore as mined carried $2 a ton in gold.

The shaft of the Gold Cliff mine is sunk from the floor of an immense open cut that had been quarried out during the operation of an old 10-stamp mill. It inclines 45° E. to the 1,600-foot level, below which it flattens to 12°.

The country rock consists of the green schists of the amphibolite schist. The ore body is mineralized schist intricately cut by quartz veins. The schist, as explained on page 43, is greatly altered and pyritized. The coarse cubical pyrite that is common in the schist is said to carry only $5 a ton in gold, as determined by assaying it separately, but inasmuch as the concentrate carries $40 a ton, the fine sulphides must be exceptionally rich. The quartz veinlets, unless they

[48] Raymond, R. W., Mineral resources of the States and Territories west of the Rocky Mountains for 1869, pp. 30-31, 1870.

contain sulphides, are poor. In other words, the gold is associated with the sulphides whether they are in the quartz veinlets or in the intervening schist.

In depth the Gold Cliff lode flattens extraordinarily, dipping as low as 7° E. on the 1,700-foot level. In stoped-out areas the remarkably fine plane hanging wall was exposed for hundreds of square feet. It stood without timbering, and this was one of the factors that aided in achieving record low operating costs.

An interesting detail exposed on the 1,700-foot level was the displacement of the lode by a steep eastward-dipping postmineral fault. It was a normal fault, which had dropped the lode 3 feet on the east side of the fault. This displacement was reported to increase southward along the strike of the fault.

The Gold Cliff lode is 3,000 feet west of the Utica vein, which crops out in the center of Angels Camp and was the mainstay of the Utica Mining Co. for many years. It intersects the Utica vein at about 2,700 feet below the collar of the Utica shaft. This intersection was the objective of many years' work in the deeper workings of the Utica mine, but the shoot of high-grade ore expected to occur there was not found. All workings of the Utica and Gold Cliff mines were inaccessible in 1924.

The Madison mine, also owned by the Utica Mining Co., was on the south extension of the Gold Cliff lode. It worked a hanging-wall ore body that rested on a thick quartz vein—the "bull ledge"—as footwall. The workings extended to a depth of 1,300 feet.

ROYAL MINE

The Royal mine is of particular interest because, although it is 14 miles west of the Mother Lode belt, the ore and its mode of occurrence are exactly like those in Amador County. Immense bodies of "gray ore," like that mined in the Fremont and Bunker Hill mines and their neighbors, were worked in it. The mine is at Hodson, Calaveras County, 3 miles northwest of Copperopolis. It is opened by an incline 1,020 feet long. A 120-stamp mill is on the property but was idle in 1924. According to Mr. F. S. Tower, lessee, the total output is $3,000,000.

The Royal vein strikes N. 40° W. and dips 30° N., although in places it is horizontal or even dips in the reverse direction. On the average the thickness of the quartz filling of the vein proper was 3 feet, though locally it was 20 feet. The vein does not occur in a diabase dike, as is sometimes stated. The main country rock consists of breccias and tuffs of the augitic greenstone series. Some lenses of black slate, undoubtedly of Mariposa age, are included in the greenstone; they strike N. 60° W. and dip 70° E.

Early operations were confined to mining the quartz vein, but later large bodies of gray ore were taken out on both foot and hanging walls. The enormous stopes, standing without timber, are results of the removal of these great bodies of gray ore. The gray ore is

identical with that of the Mother Lode belt in Amador County; it is a pyritic ankerite-sericite rock, the result of the hydrothermal alteration of the greenstone tuffs and breccia. It is reported to have milled $3.72 a ton.[49]

In the hanging-wall country rock of the Royal vein is another vein which has been stoped. The Royal vein, however, occupies the master fracture, and the rock adjacent to it on both foot and hanging wall sides was shattered, and large stockworks of gray ore were formed.

MINES ON CARSON HILL

History.—Carson Hill, near the south boundary of Calaveras County, is a prominent landmark in the Mother Lode belt, rising 1,200 feet above Stanislaus River on its south flank. The small town of Melones, formerly known as Robinson's Ferry, lies at the base of the slope, on the river.

Gold was discovered on Carson Hill in 1849, and the hill soon became noted for its pockets of high-grade ore. A vivid account of the armed struggle, fortunately bloodless, for the possession of the rich quartz vein at Carson Hill in 1851 is given by the historian Bancroft.[50] Nearly $3,000,000 in gold is reported to have been taken from the Morgan claim in less than two years, including what is said to be the largest nugget ever found in California. Rich pockets were also mined in the Carson Hill, Melones, and Stanislaus claims.[51] After the rich surface deposits had been exhausted, mining languished for many years.

About 1890 the Calaveras Consolidated Mines Co. began work on the Calaveras vein, and in 1898 the Melones Mining Co started on claims along a massive quartz vein. The ore body that the Melones Co. worked consisted of an immense mass of pyritic schist carrying $2 to $2.50 in gold, which in the course of time, from its position in the footwall of the massive quartz vein and from the need of distinguishing it from the remarkable ore body found much later on the hanging wall of the quartz vein, came to be called the footwall ore body.

An adit tunnel 4,500 feet long was driven by the Melones Mining Co. northwestward from Melones to undercut the footwall ore body, which crops out on the north flank of Carson Hill. This tunnel corresponds to the 1,100-foot level. The ore above the tunnel was mined by the glory-hole method and that below it by the shrinkage-stope method.[52] It was milled in a 100-stamp mill. The cost was lower than in any other mine on the Mother Lode belt. "The cost for 1908–1910, including all charges of every description except marketing concentrate, was $1.084 per ton."[53] By

[49] Tucker, W. B., Mines and mineral resources of Amador County, Calaveras County, Tuolumne County, p. 103, California State Min. Bur., 1915.
[50] Bancroft, H. H., California inter pocula, 1848–1856, pp. 237–240, 1888.
[51] Browne, J. R., Mineral resources of the States and Territories west of the Rocky Mountains for 1866, pp. 205, 210, 1867.
[52] Tucker, W. B., op. cit., pp. 93–96.
[53] Private report by W. B. Devereux, jr., and H. A. J. Wilkens to W. J. Loring.

1914 the total operating cost had increased to $1.60 a ton.

In that year the shaft sunk from the tunnel level had reached the 1,750-foot level. In the deeper levels a shoot of high-grade ore was discovered resting against the hanging wall of the big or Bull quartz vein. As it had been found that the Bull vein had been displaced by a number of reverse faults, the largest of which displaces the vein 100 feet, it was thought that this faulting had caused the ore shoot to jump from the footwall side to the hanging-wall side of the Bull vein. A puzzling feature of this explanation, aside from its geologic improbability, was that the hanging-wall ore was of much higher grade than that long mined from the footwall side.

The answer to the problem came with dramatic impressiveness in 1919. The old Morgan mine was purchased in 1918 for $600,000 by the Carson Hill Gold Mines (Inc.) under the management of W. J. Loring, and mining began on December 28, 1918. The first work done was the driving of the Morgan adit, on the north slope of Carson Hill. Starting from the hanging-wall side of the Bull vein, it was driven toward that vein and ran into a body of rich ore in the schist resting on the vein only a few feet from the old workings. The ore body thus intersected at less than 300 feet below its outcrop proved to be the upward extension of the high-grade ore that had originally been cut in the hanging wall of the Bull vein on the 1,600-foot level. It has come to be known as the high-grade ore body, or hanging-wall ore body. From it within the next few years was taken out more than $5,000,000.

In 1920 the property of the Melones Mining Co., which had shut down in 1918, was acquired by lease; and the subsequent development of the combined Morgan and Melones mines has been carried on by means of the Melones tunnel, which is equipped with electric traction. An interior three-compartment inclined shaft had been sunk by the Melones Mining Co. to the 3,000-foot level, and from this level at a point several hundred feet south of the main shaft a winze was sunk by the Carson Gold Mines Co. to the 4,375-foot level. The workings in the combined Morgan and Melones mines aggregate more than 15 miles.

The ore is crushed in the Morgan mill, on Stanislaus River not far from the portal of the Melones tunnel. The mill is equipped with 30 stamps of 1,250 pounds weight and has a daily capacity of 500 tons. This duty per stamp—16 tons a day—is the highest on the lode. The 100-stamp mill formerly operated by the Melones Mining Co. has been closed. The recovery of gold from the ore has averaged 90.76 per cent over a period of five and a half years. The technical details of the mining and milling methods employed have been ably presented by G. J. Young.[54]

[54] Gold mining at Carson Hill, Calif.: Eng. and Min. Jour., vol. 112, pp. 725-729, 1921.

The total output of the mines on Carson Hill is reported to be $20,000,000. The Carson Gold Mines (Inc.) produced up to July 31, 1924, $5,598,310. Its output by years is shown in the following table:

Output of Carson Gold Mines (Inc.), 1919–1924

Years	Tons milled	Yield per ton	Tailings loss per ton milled
1919	72,387	$13.71	$1.25
1920	105,156	10.40	1.14
1921	175,766	6.07	.78
1922	183,733	5.64	.59
1923	172,066	5.78	.49
1924 (to July 31)	88,869	4.66	.43

In 1924 gold-bearing rock was considered to be ore if it contained more than $3 to the ton above the 1,100-foot level, between $4 and $5 between the 1,100-foot and 3,000-foot levels, and more than $5 below the 3,000-foot level. Although some ore assaying as much as $40 to $50 a ton was found in the bottom levels (4,125 and 4,375 foot levels), the failure to find a persistent body of ore caused the mine to go into the hands of a receiver in 1925.

General geology.—The geology of Carson Hill is varied and complex—more so than that of any other locality in the Mother Lode belt. In general the rocks consist of black phyllites, augitic tuffs and breccias, chlorite schists, and amphibolite schists of several kinds, all striking northwest and dipping steeply east. As mapped by Ransome they are of Calaveras age. They have been closely folded, with the resulting production of steeply plunging folds and of abrupt variations in the degree of metamorphism from place to place. They have been intruded by small masses of gabbro and serpentine, and these igneous rocks have also been deformed; the gabbro is sheared and saussuritized, and the serpentine is in places transformed to talc schist. Widespread powerful hydrothermal alteration has put on the finishing touches, most notably by transforming much of the serpentine to great bodies of ankerite-mariposite rock.

Ore bodies.—Two veins, the Calaveras and the Bull, are of present interest. The Calaveras, the more westerly of the two, is itself auriferous and though locally known as the true "Mother Lode" has proved disappointing. The Bull vein, however, although itself not profitably gold-bearing, has associated with it a low-grade ore shoot in its footwall and a high-grade shoot in its hanging wall. Moreover, five veins, the so-called flat veins, dipping 35° E., cut the Bull vein, which dips 60° E., and contain ore at or near their intersections with it. Other less continuous veins, among them the Pink and the Stanislaus, are chiefly notable for the pockets that were found in them in early days.

The Calaveras vein, striking N. 40° W., can be traced from Stanislaus River to Carson Creek, a distance of 1½ miles. It is a stringer lode, averaging from 10 to 15 feet in width and reaching a maximum of 100 feet. Surface trenching across the lode at intervals along a stretch of 1,700 feet of its southern portion has disclosed two ore shoots of rather low grade, each 250 feet long. An open cut has been started on the more southerly of the two. The lode consists of lenses of quartz, the largest 10 feet wide, inclosed in deeply weathered pyritized schists. Ankeritized serpentine occurs locally along the lode line. The quartz filling in places is clearly of composite origin; a lens near the south end of the glory hole shows dark quartz that is cut by a network of younger veinlets of white quartz. The assay plans of the surface trenches show that the gold in the ore is dependent upon the presence of quartz.

The Calaveras vein is opened by a crosscut adit tunnel, 300 feet below its cropping, and by a drift tunnel somewhat lower. From the lower tunnel a winze was sunk and three levels turned from it. These workings are not now accessible. The upper or Calaveras adit cuts through 400 feet of augite basalt breccia and amphibolite schists and then penetrates the ore zone, which is here 100 feet wide and composed of highly crumpled fissile chloritic schists. From 2 to 5 feet of quartz occurs 60 feet west of the hanging wall of the zone. This body of quartz expands to 13 feet in the stope 30 feet above the adit. A small quantity of pyrite is present, generally in the filaments of schist inclosed in the quartz. This ore carried from $4 to $7 a tone. This ore shoot pitches 70° N.; its vertical extent is 500 feet; and on the bottom levels the ore assays from $1 to $3.20, the lower grade predominating.

In 1919-20 the Calaveras vein yielded $46,580 from 11,997 tons of ore, or $3.88 a ton.

The remarkable system of ore bodies associated with the Bull vein has no counterpart anywhere in the Mother Lode belt. On account of its great interest it will be described in some detail. The ore-bearing zone crops out on the north slope of Carson Hill, but all operations are now carried on by way of the Melones tunnel, whose portal on the south slope of the hill near Stanislaus River is at an altitude of 889 feet above sea level. In this tunnel the rocks are excellently shown. They are green or grayish green and range from massive to finely foliated. A little black slate is interbedded with them. They strike north to N. 40° W., this diversity being due to their folded condition; in places the sedimentary banding makes an angle of 20° with the foliation. Massive banded tuff at 560 feet from the portal shows under the microscope abundant fragmental augite. The marked schistose structure of the well-foliated green phyllite at 1,425 feet from the portal is shown under the microscope to be determined by finely fibrous amphibole, which has formed at the expense of augite. Not all the augite has gone over into amphibole. Representatives of these rocks are shown on the surface, among them the grayish-green

amphibolite schist occurring 200 feet north of the portal. Although this schist is well foliated, under the microscope it is seen to contain abundant pyroclastic augite, which is still on the whole remarkably intact. Obviously the rocks cut by the tunnel represent a bedded series of augitic tuffs, of which some have been more and others less thoroughly metamorphosed to amphibolite schists. The coarsest variety of the pyroclastic rocks is a breccia containing angular fragments 1 to 2 inches in diameter; it is exposed a few hundred yards north of the portal of the tunnel.

Besides the augitic tuffs and augitic amphibolite schists cut in the Melones tunnel other varieties of amphibolite schist occur on Carson Hill. There is a greenish-gray fine-grained " massive " schist, as represented by the occurrence east of the Stanislaus vein, above the railroad track. This schist proves to consist mainly of amphibole and zoisite and is so highly metamorphic that its original nature is no longer demonstrable. Another variety is the blocky amphibolite schist that occurs in the hanging wall of the Bull vein. It is excellently exposed on the surface and in the Morgan tunnel. It has been erroneously termed diorite porphyry in private geologic reports submitted to the Carson Gold Mines Co. and regarded as an intrusive mass younger than the Bull vein and a main factor in determining the position of the ore bodies. It contains numerous moderate-sized shining dark hornblende particles that might easily be interpreted as the hornblende phenocrysts of an igneous rock. However, they are clearly shown by the microscope to be paramorphs after angular fragments of pyroclastic augite. They range from fragments perfectly intact in form, though completely changed to hornblende, through those merely cracked to those that are crushed and dragged out. Fibrous amphibole is extremely abundant in the matrix, and with it occur epidote, albite, dolomite, and titanite. The rock is an amphibolite schist derived from an augitic tuff, and as direct comparison shows it is essentially like the augitic amphibolite schist occurring 200 feet north of the Melones tunnel, except that it is more highly metamorphic and that consequently all the augite once present has gone over to hornblende. In conformity with its origin as a metamorphosed bedded fragmental volcanic rock, this belt of amphibolite schist includes some chloritic slate and banded rocks.

Masses of serpentine with its alteration products, talc schist and mariposite-ankerite rock, occur as small intrusive masses in the schists. The largest mass is 200 feet in the footwall of the Bull vein on the Morgan claim; it is a green microcrystalline or finely granular antigorite serpentine. Along the lode line of the South Carolina claim another lens of serpentine occurs, but there the serpentine is entirely converted into ankerite rock with a border of talc schist. A small plug of gabbro crops out on the west, in the footwall of the South Carolina glory hole, and on the west slope of the hill is a mass of gabbro made up of saussuritic feldspar and abundant diallage. Underground serpentine and related rocks are exposed at a few points in footwall crosscuts. A thin dike of serpentine with talc schist borders occurs in the Melones tunnel 3,400 feet from the portal.

The Bull vein is remarkably constant in physical character, being composed of coarse white quartz, which is invariably ribboned with chloritic filaments. The perfect preservation of the most delicate crenulations in the schist filaments proves that the ribboning was not produced by shearing.

The foot and hanging walls are both "frozen"— that is there is no postmineral gouge on either wall. The footwall country rock is for the most part thoroughly penetrated by a network of quartz stringers, which join in unbroken continuity with the quartz of the Bull vein itself. The hanging wall of the vein shows locally a marked fluting, which rakes 75° S. The width of the vein ranges within wide limits, increasing from 18 inches on the summit of Carson Hill to more than 40 feet on the Morgan and Union claims, farther northwest. In this stretch the strike changes from N. 40° W. to S. 80° W. and back to N. 40° W.; consequently the outcrop has the form of a reverse curve, of which the southern bend is the more abrupt. As might be expected from these differences, the vein cuts across the strike of the country rocks. It does not, however, cause any disturbance of the schistosity. The angle between the course of the vein and the schistosity is generally small but locally is large, as shown in the deepest pit of the Melones glory hole or on the 1,600-foot level. (See fig. 22.)

The Bull vein has been crosscut in many places and has been adequately sampled. It rarely assays higher than $3 a ton; probably its average content is between $1 and $1.50 a ton.

The prevailing country rock in the vicinity of the ore bodies consists of thinly fissile dark-green chlorite schists. Adjacent to the Bull vein, however, the schists have been rendered pale yellow or cream-colored by the development of sericite and ankerite in them at the expense of the chlorite and other preexistent minerals. A particularly good section across the rocks in the hanging wall of the Bull vein is given by the 200-foot crosscut on the 675-foot level; the first 20 feet adjacent to the Bull vein is sericitized schist, above which in succession are blocky banded green tuffs, chloritic phyllites and slates, and finally 18 feet of faintly banded white or pale cream-colored rock which somewhat resembles a normal unaltered carbonate (dolomite) rock but is shown by the microscope to be an ankeritized rock, with minor sericite and quartz. Such ankeritized rocks are extremely abundant in the mine. They are particularly well shown on the 1,350-foot level in the south workings on the 1,100-foot flat vein, forming huge blocky exposures. There are many intermediate varieties between the massive ankerite rock and banded sericite-ankerite schist; doubtless these are all the ankeritized equivalents of the green schists seen in the Melones tunnel.

Coarse mariposite-ankerite rock having a rough foliation occurs in considerable quantity in the footwall schists of the Bull vein, undoubtedly derived from serpentinized peridotite intrusive into the schists. Exceptionally some antigorite serpentine occurs, as shown in the face of the north drift on the 1,600-flat vein on the 1,350-foot level.

A footwall spur of the Bull vein on the Morgan claim is known as the Pink vein. It can be traced for a few hundred feet southeastward.

Ore shoots occur in both the hanging wall and the footwall country rocks of the Bull vein. Two have been of great economic importance—the so-called hanging-wall ore body and the footwall ore body.

EXPLANATION

Bull vein   Green schists   Ankerite rock   Flat vein

FIGURE 22.—Plan of part of the 1,600-foot level, Melones mine, showing the divergence in strike between the Bull vein and the country rocks

The positions of these two shoots are shown on the stereogram in Plate 11, which has been accurately drawn by Thomas B. Nolan from the ample data supplied by the Carson Gold Mines Co. The outline of the footwall ore is shown by dotted lines.

The hanging-wall ore body occupies the northern of the two troughs formed in the Bull vein by the reverse curve previously described. It is the most persistent of the ore shoots, having been followed from the surface down to the 4,375-foot level. In the upper levels of the mine the shoot took the form of a bluntly terminated lens of mineralized schist, lying directly against the Bull vein. It averaged 175 feet in length and 15 feet in width, although locally its dimensions

EXPLANATION

1 Serpentine
2 Chloritic schist
3 Bull vein
4 Sericite-ankerite and chlorite-ankerite schists
5 Amphibolite schist with hornblende porphyroblasts
6 Massive ankerite rock
7 Ore

0 100 200 300 400 500 FEET

FIGURE 23.—Partial section through the Melones mine along the line of the lower winze on the 1,100-flat vein

were much greater. The shoot was not restricted to rock of any one type, equally good ore occurring in both chloritic and sericite-ankerite schist. The ore consists of sericitic schist composed largely of ankerite, with less pyrite, sericite, quartz, and albite. The pyrite is particularly abundant in high-grade portions, where it is accompanied by chalcopyrite, galena, tetrahedrite, native gold, and locally petzite. Molybdenite was fairly common in parts of the ore body; ore containing it was termed "graphitic" and was looked on with favor as being above average grade. Although the bulk of the ore is a pyritic sericite-ankerite schist, or in part, where the chlorite has not been all sericitized, a pyritic chlorite-ankerite schist, locally ores

of different kinds occur. One of these was the ore on the hanging wall of the hanging-wall ore body of the 1,750-foot level, which might be called a pyritic jasperoid; microscopically it proves to consist of a fine-grained aggregate of albite, which predominates, and ankerite. The ore from the 3,200-foot stope was a banded pyritic jasperoid, whose marked striping was due to the alternation of thin chlorite layers with layers composed of albite and ankerite.

The ore body kept its average dimensions down to the 3,000-foot level but gradually decreased in grade. Below that level the ore almost pinched out, coincident with the disappearance of the sharp curve in the Bull vein. The shoot widened below the pinch and has been mined down to the 4,375-foot level. Below the pinch, however, the shoot is separated from the Bull vein by 1 to 40 feet of leanly auriferous schist. More quartz occurs in the ore of the deeper levels than in the higher ore, but whether this greater quartz content is a mere local variation, as appears most probable, or marks the beginning of a permanent change in the character of the ore can not be predicted from any facts disclosed at the mine or from any principles known to geologic science. The distribution of the ore in the shoot or ore zone on the 4,125-foot and lower levels proved to be highly irregular. Some high-grade ore, carrying $50 a ton, was discovered, but no continuous body had been found in 1924.

The footwall ore body extended from the surface to the 2,170-foot level and had been completely extracted before 1924. The ore consisted of pyritized schist full of quartz stringers and was of low grade, ranging from $2 to $2.50 a ton. The shoot was mined by the glory-hole method down to the 1,100-foot level. A width of about 40 feet was extracted.

Ore has been stoped from five other veins—the so-called flat veins. The veins strike 30° to 40° more westerly than the Bull vein, and they dip much less steeply, generally 35° N. but in places flattening to 20°. A fine planar hanging wall is distinctive, as is also the presence of a few inches of postmineral gouge on both walls. The Bull vein is faulted reversely by the flat veins, the maximum displacement being 120 feet at the intersection of the Bull vein and the flat vein known as the 1,100-flat vein. (See fig. 23.) The displacement of the country rocks by the flat vein is shown in Figure 23.

The flat veins contain ore shoots only near their intersections with the Bull veins. (See pl. 11.) Of the five flat veins that have been mined the second lowest, or 1,100-flat vein, has been the most productive. The average stope length on this vein exceeded 300 feet from the 1,100-foot level to the 1,600-foot level. The shoot pinches and swells abruptly, ranging in thickness from a few inches to more than 30 feet. This maximum thickness, however, includes a large portion of mineralized wall rock as well as the vein filling proper. The ore is of three kinds, which grade into one another. They are (1) coarse white quartz carrying sparse pyrite and containing numerous vugs lined with flat rhombohedral crystals of ankerite; (2) fis e filling consisting of a rubble of angular blocks of mineralized wall rock cemented by quartz veinlets; (3) mineralized wall rock consisting of blocky ankerite or ankerite schist, carrying abundant large well-formed pyrite crystals. Spots of high-grade material occur locally near the footwall of the veins and contain in addition to the pyrite small quantities of galena and native gold. As this high-grade ore gave some trouble in extraction at the mill, it was thought that the galena-like mineral might be the megascopically similar lead telluride, altaite, but laboratory tests failed to disclose tellurium in the samples that were collected to test this supposition.

Ankerite was the predominant constituent in all this ore, and this fact together with the abundance of huge cubes of pyrite made it unlike any other Mother Lode ore. It averaged from $6 to $10 a ton and not only supplied the bulk of the material mined during 1924 but on account of the ease with which it could be mined by the use of scrapers it supplied ore at far lower cost than the hanging-wall ore body.

Apparently related to the flat veins in origin is a vein as yet exposed only on the 675-foot and 985-foot levels. It is unusual in that it dips 45° W. Like the flat veins it swells and pinches abruptly and is accompanied by large cubes of pyrite in the adjacent wall rock. No ore has yet been found in this vein.

The concentrates from all ore, both from the flat veins and the hanging-wall ore body, form 3 per cent of the ore milled; they carry $40 a ton.

NORWEGIAN MINE

By Thomas B. Nolan

The Norwegian mine is in Tuolumne County on the south bank of Stanislaus River, half a mile southeast of the town of Melones. The mine has been worked at intervals for many years; the production has come almost entirely from pockets rich in native gold

and tellurium minerals. In August, 1924, the Norwegian was being operated by two lessees.

The mine is developed by an inclined shaft 550 feet deep, the lower 250 feet of which, however, was flooded at the time of visit. Several levels have been driven, but only the two lower ones are of any present extent. The high-grade ore mined is hand crushed and concentrated.

The shaft is sunk on the Norwegian vein, which strikes N. 12° W. and dips 55° E. The vein crops out in amphibolite schist near the western contact of the schist with a tongue of the Calaveras formation that comes in from the north. The Calaveras is here composed of black slates and siliceous beds that show a characteristic fluting. The amphibolite schist in the few unaltered exposures found is a fine-grained gray-green schist. Pyritization and ankeritization of the schist is the rule and is notably intense for a distance of 50 feet in the hanging wall of the vein. A discontinuous barren or "bull" quartz vein marks the eastern boundary of this intense mineralization and makes an acute angle with the Norwegian vein.

The Norwegian vein is not being worked at the present time but, according to Ransome,[55] the vein was from 4 to 20 inches wide and frozen to its walls. The vein minerals were quartz, calcite, dolomite, and a little albite and chlorite, with pyrite and smaller quantities of native gold, bornite, galena, chalcopyrite, petzite, hessite, coloradoite, and other tellurides.

A vein 12 feet in the footwall of the Norwegian vein known as the footwall vein, was being mined in 1924. On the 300-foot level it consisted of 6 inches of quartz and ankerite. The strike and dip of the country rocks differ from those of the vein, and as a result different types of rock are found in the walls as the vein is followed. On the 300-foot level a belt of black slate was intersected by the vein, and a very high grade pocket was developed. This with previous similar experiences has led the lessees to form the empirical rule that rich pockets occur at intersections of veins with the black slate belt.

The ore from such pockets is composed of quartz and ankerite, rather abundant native gold, galena, and petzite, and 5 to 10 per cent of pyrite. Other tellurides are probably present, but the only specimen collected at the mine proved to be petzite as verified chemically. The concentrates from the pockets are worth from $3.50 to $4.50 a pound

Both the Norwegian and footwall veins represent minor fissures, along which there was little or no displacement. They are probably sympathetic to some main fracture, possibly that occupied by the barren quartz vein previously described. If this is true, the intersection of the barren vein and the Norwegian vein might be well worth prospecting.

55 Ransome, F. L., U. S. Geol. Survey Geol. Atlas, Mother Lode District folio (No. 63), p. 9, 1900.

CHILENA MINE

By Thomas B. Nolan

The Chilena mine is near the summit of Jackass Hill, in Tuolumne County, three-quarters of a mile northwest of Tuttletown. Jackass Hill contained many rich pockets that were diligently prospected in the early day of mining in California, but since the exhaustion of the deposits near the surface the claims have lain idle except for surface prospecting by individuals. In 1922 the Chilena together with adjoining claims was optioned by the Nevada Wonder Mining Co., and a fair amount of development work was done on it. The results were disappointing, however, particularly on the lower level, and the project was abandoned in the fall of 1923. The property was then leased to the Mark Twain Mining Co., which intends to stope some ore developed on the upper level by the former company.

An inclined shaft and a 150-foot and a 450-foot level were driven by the Nevada Co. The lower level was flooded when the mine was visited. A small Huntington mill treats the ore.

The country rocks are amphibolite schists, a fine-grained greenish type being predominant. Some coarser-grained facies also occur, one of which contains hornblende porphyroblasts.

The ore of the mine is obtained from a zone of intensely metasomatically altered schist, 4 to 9 feet wide, on the 150-foot level. The strike of the zone averages N. 40° W. but ranges between N. 25° W. and N. 50° W. The plan of the zone therefore consists of a series of arcs alternately concave and convex. The dip averages 68° E. but locally steepens to 85°.

The mineralized zone is marked by the occurrence of quartz, albite, sericite, and pyrite, with minor amounts of a carbonate mineral. The resulting rocks are cream-colored and contrast strongly with the normal greenish schists. The walls are not sharply defined; in places as much as a foot of transition rock separates the ore zone from the surrounding schists.

An ore shoot 160 feet long has been developed in the zone. According to the superintendent, Mr. L. L. Coffer, the shoot averages $9.20 a ton. The distribution of the gold is somewhat erratic, portions of rather high-grade ore alternating with poorer material. Telluride minerals are reported from some of the high-grade pockets. The sulphides, essentially all pyrite, form 15 per cent of the ore and assay $20 a ton, thus furnishing a considerable portion of the yield.

The few mine workings accessible and the poor surface croppings around the mine do not warrant a definite statement of the nature of the ore zone. Probably it represents a shear zone in amphibolite schists, which was later metasomatically altered by ascending mineralizing solutions carrying gold.

DUTCH-SWEENEY AND APP MINES

The Dutch-Sweeney and App mines are at Quartz, 2 miles southwest of Jamestown. The Dutch and Sweeney claims together comprise 3,412 feet along the Mother Lode system. The Dutch claim was located in 1852, but it was not extensively developed until 1893. It is believed to have produced under the Dutch Gold Mining & Milling Co. $2,000,000 gross. The plant consisted of a 20-stamp mill and the shaft extended to the 1,800-foot level.

There are three veins on the property—the Heslep or hanging-wall vein; the middle vein, a broad lode of ankerite stringered with quartz but only locally of commercial grade; and the App or footwall vein.

The early work was on a rich shoot in the footwall vein. This shoot was 490 feet long near the surface and ranged in width from 6 to 30 feet. It proved to be profitable down to the 400-foot level, averaging $10 a ton. Attention was then centered on the Heslep vein in the hanging wall, which averaged about $4 a ton. In 1906 the company went into receivership, and eventually the Dutch property was sold. In 1909 the Dutch claim was consolidated with the Sweeney, adjoining it on the north. The acquisition of this ground was particularly desirable on account of the northward rake of the ore shoots. In 1909 the Dutch Consolidated Gold Mining Co. began operations. It enlarged the mill to 40 stamps in 1912 and milled about 200,000 tons of ore, which yielded slightly less than $3 a ton.

In 1917 the property was taken under option by the Pacific Coast Gold Mines Corporation, of which W. J. Loring was president and manager. At the same time the App mine, adjoining the Dutch on the south, was also acquired; the App was prospectively valuable on account of the northward rake of the App shoot. A new mill was built and started operating in January, 1919. It ran until April, 1920, and milled 54,432 tons of ore, from which was obtained $41,994 in bullion and 2,855 tons of concentrate, containing $133,687, a total of $175,681 or $3.23 per ton milled. An extraction of 90.2 per cent was made.

The new exploratory work done before the mine was shut down, in 1920, amounted to 16,000 feet. The shaft was deepened from the 1,800 to the 2,300 foot level. The Heslep shoot on the 1,500-foot level, which proved to be the best level, was 400 feet long and 10 to 20 feet thick and averaged $4 a ton, but on the 2,300-foot level it had shortened to 40 feet. The App or footwall shoot was also prospected. The general result of the exploration was to show, according to O. H. Hershey, "that the ore shoots vary constantly from level to level in size, shape, and gold content." Because of the unfavorable postwar conditions the average gold content of the ore, which was under $4 a ton, proved to be too small to yield a profit. The mine was therefore shut down in 1920 and has remained idle since then.

EAGLE SHAWMUT MINE

The Eagle Shawmut mine is on Woods Creek, 2 miles northwest of Jacksonville, in Tuolumne County. It is the property of the Eagle Shawmut Mining Co. Not much work had been done before 1896, when this company combined the Eagle and Shawmut claims. The mine was operated under this management until 1916, when the Tonopah Belmont Development Co., through a subsidiary known as the Belmont Shawmut Mining Co., obtained control under an option to purchase. Considerable exploration was done, but in November, 1923, work was stopped because of the low grade of the ore developed. During this period the mine was known as the Belmont Shawmut. An excellent description of the mining and milling methods employed is given by Parsons.[66] In 1924 the original company had resumed operations, and the older name was in use.

The mine is developed by a crosscut adit driven 1,100 feet through the footwall to the lode and connecting with an inclined shaft 2,163 feet deep. From this internal shaft have been driven at various intervals 19 levels. From the bottom of the nineteenth level a 900-foot winze has been sunk and two levels turned from it. The total vertical depth attained below the outcrop is 3,000 feet.

One of the innovations on Mother Lode practice is the use of a storage-battery locomotive underground, on the nineteenth level. As a labor-saving device it has proved eminently satisfactory; hardly had it been put into operation when it made possible the removal of 21 men from the pay roll.

A 70-stamp mill treats the ore. Flotation equipment had been installed by the Tonopah Belmont Development Co., but on resumption of control by the owners it was discarded. About 300 tons of concentrate carrying $60 a ton is produced monthly and is sent to the Selby smelter for treatment. The total output to the end of 1924 was about $5,000,000, which was obtained from 1,750,000 tons of ore.

The lode system worked in the Eagle Shawmut mine lies at the contact of the Mariposa slate and the Calaveras formation, which here contains much green schist. The southern part of the system is marked by the immense outcrops of white quartz that extend from the summit of the ridge 1,400 feet south to Blue Gulch.

The Mariposa slates are interbedded with innumerable thin layers of sandstone averaging 2 inches in thickness. In strike they range from N. 20° W. to N. 45° W. within short distances. The cause of this diversity of strike is excellently shown in the main tunnel, which is the best place along the Mother Lode to see the close folding that has affected the Mariposa beds. At 200 feet from the portal the bottom of a perfect synclinal fold pitching 60° S. is shown; the axial plane of the fold strikes N. 32° W. and dips 80° E., and the fold causes the beds on its limbs to diverge 30° in strike. Another similar fold is well shown 500 feet farther in. On the surface, near the Bull vein, the Mariposa slate incloses a belt of volcanic rock that includes an amygdaloid flow and a breccia composed of angular fragments of amygdaloid as much as 6 inches in diameter. At Blue Gulch the amygdaloid is 100 feet thick; it is a basalt crowded with augite phenocrysts and amygdules of calcite and chlorite. This belt of amygdaloid and breccia was mapped by Ransome as metadiabase, but as the rocks are neither intrusive nor of diabasic texture that name can not be retained. Near the upper contact of the slate with the volcanic belt occur some lenses of limestone 2 to 3 feet thick, and near the old loading station there is a thicker mass, resting on black slate. Some thin lenses of conglomerate also occur in the slate below the amygdaloid breccia.

Between the Mariposa slate and the gray-green chlorite schists associated with the Calaveras formation on the east, is a narrow belt of pyroxenite or hornblendite or an allied hornblende-pyroxene peridotite. In this report it will be called the pyroxenite dike. It has been powerfully sheared and converted into multiform chlorite schists, and subsequently these schists have been extensively ankeritized and silicified. These chlorite schists, some of which to the unaided eye are indistinguishable from serpentine schist, and their ankeritized equivalents are among the chief rocks in the ore zone. One of the best exposures is in the old prospect tunnel in the northwest corner of the Bella Union claim. The rock here is an exceedingly crumbly lustrous green schist, which microscopically is found to consist of broken green amphibole fibers in a matrix of chlorite. In this crumbly schist are some partly sheared lenses or horses of hornblendite, composed of thick hornblende columns 1½ inches long. At places in the mine, as for example in the hanging wall of the Shawmut vein on the fifth level and in the footwall crosscut on the ninth level, chlorite schists occur that resemble serpentine schists; that on the ninth level, moreover, has a pseudoporphyritic texture much like that of the serpentinized peridotites, but the apparent phenocrysts instead of being bastite are chlorite and are embedded in a chlorite matrix. The only obvious physical difference between them and serpentine is that they are softer—in fact, they can be scratched with the finger nail.

The Calaveras rocks, dipping 70°–80° E., make an angle of 20° or less with the strike of the ore zone. They consist of black slates, which differ from the Mariposa slate in not containing interlayered sand-

66 Parsons, A. B., The mine and mill of the Belmont Shawmut Mining Co.: Min. and Sci. Press, vol. 121, pp. 619–624, 659–664, 1920.

stone, and siliceous slates and quartzites that are so fine grained as to verge on cherts; and they are asso-

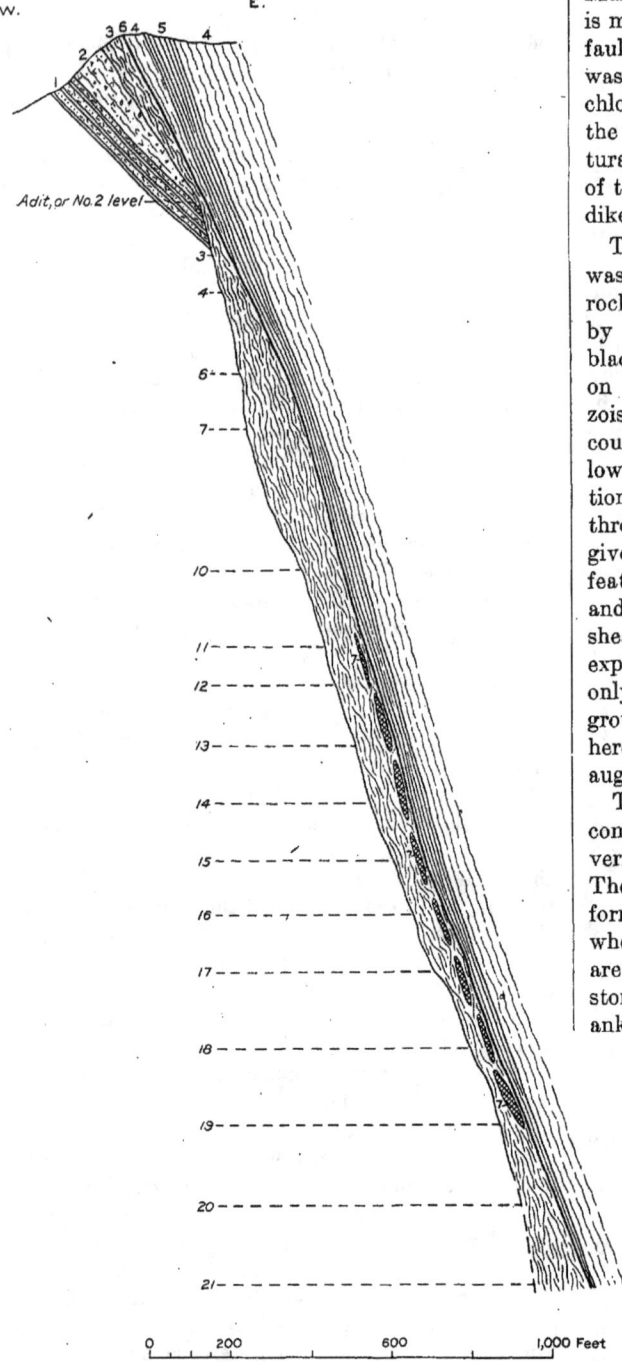

FIGURE 24.—Vertical section through the 610-foot winze, Eagle Shawmut mine. 1, Mariposa slate with interbedded sandstone or graywacke; 2, augite basalt breccia and augite porphyry; 3, sheared pyroxenite (chlorite schist), including ankeritized rock and quartz; 4, chlorite-zoisite schist; 5, Calaveras slate; 6, Shawmut vein; 7, lenses of sulphide ore (diagrammatic)

ciated with gray-green chlorite-zoisite schists, which are firm, tough rocks.

Structurally the Calaveras rocks, although older than the Mariposa slate, rest discordantly on the Mariposa rocks. This superposition of the Calaveras is manifestly the result of a reverse fault. Along this fault contact the pyroxenite dike was intruded and was afterward, as already explained, sheared to chlorite schist of various facies. The geologic map of the adit level (pl. 12) illustrates some of these structural relations, especially the truncation of the belts of the Calaveras rocks as they abut on the pyroxenite dike.

The reverse fault along which the pyroxenite dike was intruded cuts also the dip of the hanging-wall rocks. The angular divergence is small but is shown by the presence in the southern part of the mine of black slate above the ore on the lower levels, whereas on the upper levels the hard gray-green chlorite-zoisite schist occupies this position. The footwall country rocks are not exposed by any crosscuts on the lower levels, and consequently the extent of truncation of those rocks is unknown. A vertical section through the 610-foot winze is shown in Figure 24 and gives an excellent epitome of the essential geologic features of the mine. The belt of augite basalt breccia and augite porphyry shown in the figure below the sheared pyroxenite dike is practically unsheared in its exposures on the surface, but on the adit level, the only place where it is certainly recognizable underground, it is intensely sheared, and it is recognizable here only because it contains a few lenses or horses of augite porphyry that are less thoroughly schistose.

Talc schist and mariposite are supposed to be common in the mine, but all specimens collected to verify this belief proved to contain only chlorite. The siliceous slate and quartzite in the Calaveras formation have attracted much attention at the mine, where they are known as silicified limestone; but they are, so to speak, exactly the reverse of silicified limestone, for they are commonly more or less replaced by ankerite, and in many of them the sharp rhombs of ankerite can be seen with the unaided eye lying scattered through the siliceous matrix.

The ore zone comprises the sheared pyroxenite dike and the adjacent Calaveras rocks. The dike and the chloritic equivalents into which it was transformed have been tremendously ankeritized and are locally cut by a network of quartz veins and replaced by quartz. Such a local development of quartz has formed the Bull vein.

The ore mined is of two general kinds; one, known at the mine by a variety of names, is largely siliceous, and the other, called "sulphide ore," is ankeritic and carries much pyrite.

Ore of the first kind is the result of the filling by quartz of minor fractures in the Calaveras rocks. The fractures are planes of little or no movement, as gouge is rare and brecciation is absent. A fine planar hanging wall dipping 65°–70° E. is characteristic. Three veins of this kind have been developed on the adit level—the middle, east, and Shawmut winze veins. They are comparatively narrow and of low grade. For these reasons no effort has been made to explore them on the lower levels.

Two other varieties of siliceous ore occur. Some large lenses of slate included in the altered pyroxenite contain enough gold to be ore. The "slate" ore body and has been followed down to the lowest level. It is persistently of low grade, rarely assaying above $5 a ton, and between the sixth and tenth levels it averaged less than $2 a ton. On the upper levels the Shawmut vein was stoped at intervals along 1,200 feet, but the higher costs now prevailing have reduced the length of the profitable portion to 300 feet or less.

The "sulphide ore" is found between the tenth and twentieth levels and is the present mainstay of the mine. It consists of compact fine-grained blackish-

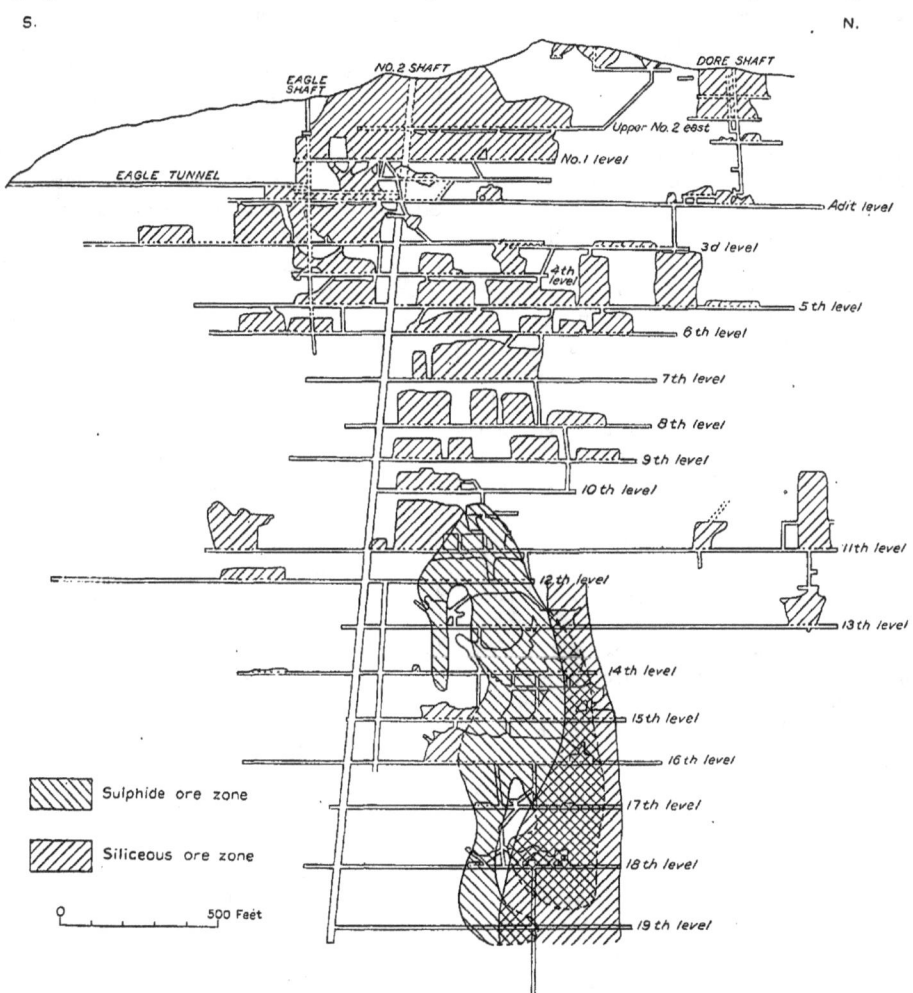

FIGURE 25.—Longitudinal section through the Eagle Shawmut mine, showing the distribution of the ore by kinds

on the eighteenth level is of this kind. Ore also occurs near or along the contact of the Calaveras rocks and the pyroxenite dike. Here, however, in contrast with the conditions at the three veins previously mentioned, a gouge has been formed on the footwall of the ore body. The gouge generally contains quartz and in places carries sufficient gold to be ore.

Of these siliceous ore bodies the one at or near the contact of the Calaveras rocks and the dike is the most persistent. This vein is known as the Shawmut green chlorite schist that has been replaced by pyrite, arsenopyrite, and varying amounts of ankerite. The ore occurs in lenses that overlap one another in a zone 400 to 500 feet long and attain a maximum width of 30 feet.

Their positions in the ore zone are shown somewhat diagrammatically in Figure 24. The gold content ranges from $10 to $16 a ton. The sulphides form as much as 15 per cent of the ore, and the concentrate that they yield averages $60 a ton. Sulphide ores of two kinds, "white" and "black," are recognized in the mine. Both are of the same origin and represent merely different degrees in the thoroughness of replacement; the "white" ore has a higher ankerite content, and the "black" ore contains more unreplaced chlorite schist. The footwall rock of the ore lenses is the banded ankerite rock, and the hanging-wall rocks are Calaveras slates and siliceous beds. A well-defined wall, along which gouge occurs locally, is present either at the contact of the sulphide ore with the slates or in the slates not far from the contact. The intervening slate is mineralized and constitutes ore, carrying $4 to $6 a ton.

The winze from the nineteenth level has been sunk (780 feet at the time of visit) in the locus of the sulphide ore lenses, following down under the well-defined hanging wall. It is reported that all the way down the assays ran between $3 and $4. The distribution of the siliceous and sulphide ore throughout the mine is shown in Figure 25.

The sequence of events leading to the formation of the ore bodies is as follows: (1) A basic plutonic rock (hornblendite, pyroxenite, or peridotite) was intruded along a reverse fault fissure between the Calaveras slates and the associated gray-green schists on the east and the Mariposa slate and associated volcanic rocks on the west; (2) renewed movement on the fault highly sheared the peridotite and converted it into various chlorite schists, and at the same time auxiliary fractures were formed in the hanging-wall rocks; (3) ascending mineralizing solutions altered the compact chlorite schist to bodies of sulphide ore and the other schist to quartz-ankerite rock, in which the quartz is locally so abundant as to form thick masses, such as the Bull vein. Concomitant with this action the Shawmut vein was formed near or along the hanging wall of the sheared pyroxenite dike, and the subordinate or auxiliary fractures in the hanging-wall country rock also became filled with low-grade gold-bearing quartz.

CLIO MINE

The Clio mine, owned by the Clio Vindicator Mines (Inc.) is on the north bank of Tuolumne River half a mile east of Jacksonville. The mine is an old one and had been idle for many years when the present company obtained control in 1917. Active work was not started, however, until 1920. The mine is developed by an inclined shaft, 660 feet deep, which dips 68° E. and from which five levels have been driven. Both men and supplies commonly enter the mine by means of the main adit, which corresponds to the 218-foot level.

Milling practice is totally different from the usual procedure on the Mother Lode. The ore is crushed directly in cyanide solution by a 10-stamp mill, no amalgamation or concentration being attempted. An extraction of 90 to 94 per cent on ore carrying $3.50 a ton is claimed. The gold bullion thus obtained is worth $12.78 an ounce. According to the manager, Mr. H. Hauter, an output of $165,000 by former companies is shown by mint records, and total output of $200,000 is probable. In September, 1924, the mine was producing at the rate of $5,000 a month.

The ore bodies of the Clio are lenses that occur in a zone of mineralized members of the Calaveras formation, whose gold content is higher than that of the rest of the ore zone. In addition to the rocks of the Calaveras formation greenstone, serpentine, and dikes of several types occur in the vicinity of the mine.

In the footwall of the ore zone the most extensive exposures are basaltic greenstone—breccias, tuffs, and lavas of the Jurassic period of volcanism. A belt 900 feet wide is exposed along the river west of the mine. Near the base of this series is 15 feet of ellipsoidal basalt, showing well-developed variolitic structure. Similar structure was also noted higher in the series. Near the mine office the belt includes sheets of amygdaloidal basalt. To the northwest, on the adjoining Kershaw claim, only 300 feet of greenstone is present, tuffs and breccias predominating. This apparent thinning is very probably due to the oblique trend of the ore zone relative to the greenstones, which here form the immediate footwall. Black Mariposa slate is exposed beneath the greenstones.

The hanging-wall country rock consists of black and green slates of the Calaveras formation. A few siliceous beds are included, also lenses of green chloritic schist that represent metamorphosed tuffs and breccias. One of the lenses on the Kershaw claim is 150 feet thick. In surface exposures these schists are usually so altered by weathering that no precise petrographic name can be applied to them.

In the immediate footwall of the ore zone on the Clio claim is a body of serpentine. "Horses" (less altered masses) of the peridotite from which the serpentine was derived are rather common. These horses show a pseudoporphyritic texture, due to the presence of large plates of serpentinized pyroxene. At the shaft 80 feet of serpentine is exposed, and the thickness increases somewhat to the south. To the north the serpentine thins abruptly, and near the center of the Kershaw claim it is absent.

A 15-foot dike of gabbro rich in hornblende occurs in the greenstone. It is considerably sheared. Other small dikes of similar composition occur in the serpentine areas. Also a hornblende lamprophyre dike, 5 feet thick, is exposed on the lode line of the Kershaw claim, 100 feet northwest of the Clio end line. Granular white dikes occur just under the summit of the hill N. 8° E. of the Clio shaft. There are light-colored albitites—the soda syenite granophyre of the Mother Lode folio—which have been considerably pyritized and whose original dark mineral has been altered to minute spherulites of the blue amphibole riebeckite. A similar dike farther north is cut by numerous quartz veinlets

The geologic relations seen at the surface continue underground, but in addition 5 to 10 feet of gouge, dipping 68° E., is found to separate the serpentine from the ore zone. The gouge is not sufficiently resistant to erosion to crop out. The ore bodies are lenses in the Calaveras formation that have been sufficiently mineralized by pyrite and quartz, both carrying small quantities of gold, to constitute ore. Sharp walls are therefore absent. This zone of mineralized country rock is 80 to 100 feet thick and includes black slates, in places delicately interlaminated with green slates, hard gray-green chloritic schists, and siliceous nearly quartzitic beds. Quartz stringers were so numerous in two of the lenses that they were locally called "quartz shoots." A "bull" quartz vein occurs in places along the contact of the gouge and the ore zone. It does not constitute ore.

At the north end of the 500-foot level a quantity of specimen ore was found, associated with a nearly flat transverse quartz vein 2 to 3 inches thick. The gold occurs in quartz and calcite together with unoxidized pyrite and is clearly primary.

The succession of geologic events at the Clio mine was as follows: First reverse faulting of unknown displacement thrust the Calaveras formation with its intercalated schist layers upon the younger greenstone series. Along this fault at a somewhat later time was intruded a dike of peridotite. This peridotite in turn was intruded by dikes of hornblende gabbro. The dikes of hornblende lamprophyre and albitite were injected possibly during this same general period. Serpentinization of the peridotite and shearing of both it and the dikes was accompanied or closely followed by recurrence of movement along the reverse fault, resulting in the formation of the gouge. Still later mineralizing solutions, ascending along the hanging wall of the gouge, formed the lenses of ore now being worked. Figure 26 shows the geologic relations. A section drawn farther north would show a reduced thickness of serpentine, and one farther south a greater thickness.

MARIPOSA ESTATE

History and general features.—The southern portion of the Mother Lode belt, that portion within Mariposa County, extending from Merced River to the termination of the belt by the granodiorite 2 miles south of the town of Mariposa, is included within the Mariposa estate. The estate is a grant of 44,387 acres (70 square miles) made by the Mexican Government to Juan B. Alvarado when California was still part of Mexico. It was a floating grant—that is, one that gave the grantee the right to fix its boundaries anywhere within a specified larger area. It was bought in 1847 by Col. J. C. Frémont, who "floated" its boundaries to include the gold discoveries, and after much litigation United States patent was issued to him in 1856. Further litigation ensued owing to the fact that under

Mexican laws title to mineral rights was severed from the title to the land, but in 1859 the courts confirmed Frémont in the mineral rights. Although there have been a number of changes of ownership the estate has remained intact to the present day.[57]

The mineral possibilities of the estate aroused unbounded optimism in all the early writers. Whitney, in fact, wrote that "the quantity of material which can be mined may, without exaggeration, be termed inexhaustible." The most celebrated mines are the Princeton, Josephine, and Pine Tree. To facilitate working the Josephine and Pine Tree mines, which are on the east slope of a ravine known as Hells Hollow, tributary to Merced River, a tramway was constructed extending to the river, where the ore was treated in a mill run by water power—the historic Benton mill. The mill is no more, but the site is on

FIGURE 26.—Section through the Clio shaft. 1, Greenstone; 2, gabbro-hornblendite; 3, serpentine; 4, gouge ; 5, ore zone; 6, Calaveras formation

the south side of the river at Bagby, on the Yosemite Valley Railroad.

By 1863 this noble estate, to use Silliman's words, had come into a disastrous position, and it has never recovered since. From the Benton mill the River tunnel was begun in June, 1874, and driven to undercut the Pine Tree and Josephine veins.[58] It is 3,380 feet long, and its average trend is S. 34° E.

The ores of the Princeton, Josephine, and the Pine Tree veins decreased abruptly in value after a moderate depth had been attained. In this respect they appear to differ markedly from many Mother Lode veins, but facts are not at hand to discuss the matter adequately. The ore from the outcrop of the Princeton mine is said to have yielded as high as $75 a ton for a short time, doubtless owing in the main to mechanical enrichment by gold liberated from the oxidized sulphides. In 1862 and 1863 the average yield was $16.50

[57] Browne, J. R., Mineral resources of the States and Territories west of the Rocky Mountains for 1867, pp. 21-30, 1868. Mariposa estate, its past, present, and future, 62 pp., New York, 1868. Frémont, J. C., and Billings, Frederick, The Mariposa estate, 63 pp., London, 1861 (contains reports on the estate by J. D. Whitney, J. Adelberg, F. Claudet, and T. W. Park).

[58] Rolker, C. M., The late operations of the Mariposa estate: Am. Inst. Min. Eng. Trans., vol. 6, pp. 145-164, 1879.

a ton, but in 1864, when the depth reached was about 500 feet, the yield fell abruptly to $6 a ton. Later work at greater depth appears not to have disclosed ore of the grade found in the upper levels. The ore from the upper levels of the Josephine yielded $22 a ton,[59] and that from the Pine Tree yielded $26 a ton down to a depth of 100 feet.[60] It is estimated that on account of the poor metallurgical methods employed a large part of the gold was lost—as much as 70 per cent. At a moderate depth both mines yielded ore whose recovered value was $6 to $8 a ton.

In 1924 the only operations in progress on the Mariposa estate were at the Queen Specimen mine, near Bagby, and at the Princeton mine, half a mile west of Mount Bullion.

Queen Specimen mine.—The Queen Specimen mine is three-quarters of a mile southeast of Bagby. During the three years 1922–1924 operations were carried on at this mine by W. M. Deaner, under a lease from the Mariposa Commercial & Mining Co. A 10-stamp mill was built on the property. It is reported that 3,000 tons of ore was milled that yielded $4 a ton and tailings that carried $2.50 a ton.

The mine is opened by a crosscut tunnel 500 feet long, whose portal is 340 feet above Merced River. The tunnel reaches the vein from the hanging-wall side. It first cuts through 130 feet of serpentine, which incloses small masses of black slate, and then goes through 370 feet of black slate, dipping 50° E. but steepening near the vein, which dips 60°–70° E. The tunnel did not intersect the ore shoot in the vein, but on drifting south the shoot was found. It ranges in thickness from a few feet to 12 feet and carries from $4 to $12 a ton. The drift on the vein is 400 feet long. A footwall crosscut shows that the vein is underlain by (1) ankerite rock netted with quartz veinlets, 15 feet thick; (2) talc schist, 30 feet thick; (3) "diorite," as it is locally called, in part sheared, hydrothermally altered, and pyritized and in part hard and massive, 48 feet thick; and (4) black slate.

Work of former lessees above the tunnel level shows that they stoped out 1 to 1½ feet of ore on the footwall of the vein. The quartz in these old workings, which continue up to the surface, is in places stained blue by azurite, doubtless derived from the oxidation of tetrahedrite, and carries pyrite and minor arsenopyrite.

Princeton mine.—The Princeton mine is on the Mariposa estate 4 miles northwest of Mariposa and half a mile west of Mount Bullion, on the Bagby-Mariposa road. It is one of the oldest mines of the Mother Lode belt, having been first worked in 1852. Rich ore was extracted from its upper levels. In the early years it was thus described:[61] "The Princeton vein, a very rich and extensively worked vein, * * * contains $35 rock in all parts," but in 1864, at a depth of 500

feet, the yield dropped abruptly to less than $6 a ton,[62] and the mine was closed shortly thereafter.

In 1863 the Mariposa grant had passed from General Frémont to the Mariposa Land & Mining Co. About 1900 the mine was again opened, this time by the Mariposa Commercial & Mining Co., which was then in control of the estate. An inclined shaft 1,660 feet deep was sunk, attaining a vertical depth of 1,250 feet, and eight levels were turned from it. The rather low grade ore developed was stoped from the 1,200-foot level to the surface. Operations by the company ceased about 1914, and by 1924 all surface equipment had been removed from the vicinity of the old shaft.

The total output of the old Princeton mine is $5,000,000, nearly three-quarters of which was obtained before 1865. Raymond[63] says that from the evidence of stope maps the mine produced between $4,000,000 and $5,000,000 from a single large chimney of ore that was worked to a depth of some 600 feet. Since the reopening in 1900 it has produced another $1,000,000.

Late in 1921 the mine came under the control of the Princeton Gold Mines Co. An inclined shaft was sunk 2,000 feet south of the old shaft, and two levels were driven at 150 and 300 feet. Late in 1924 plans were announced that included deepening the old shaft below the 1,600-foot level.[64] A 20-stamp mill treats the present output.

The vein in the new workings resembles closely the veins of Amador County in that the ore occurs in a quartz vein that occupies a reverse fault fissure in the Mariposa black slate. The vein crops out near the western boundary of the wide belt of slates that passes through the town of Mariposa. The slate strikes N. 35° W. and dips 70° E. This belt is terminated 2 miles south of the town by the typical normal granodiorite of the Sierra Nevada. Thin graywacke beds are included in the slate, and 250 feet west of the shaft a lens of massive augitic graywacke is exposed. The vein croppings are well shown in surface workings, 3 to 4 feet of quartz dipping 45° E. being exposed in practically all the pits. In a prospect hole about 800 feet S. 17° W. of the shaft a sheared porphyry dike about 10 feet thick has been cut. The porphyry is light colored and contains phenocrysts of feldspar. A heavy growth of greasewood, manzanita, and scrub oak makes difficult a close study of the surface geology of the area surrounding the mine. There is some question as to whether the vein now being worked is the southward extension of the one worked from the old shaft, or a footwall branch, or a separate vein situated in the footwall of that vein. From the poor exposures available it is impossible to decide.

The vein dips 40°–45° E.—much less steeply than the inclosing slates. The divergence between the dip

[59] Raymond, R. W., Mineral resources of the States and Territories west of the Rocky Mountains for 1868. Gives a stope map showing thickness and yield per ton.
[60] Idem, p. 14.
[61] The Mariposa estate, p. 12, London, 1861.

[62] Ashburner, William, Geological formation of Pacific slope: Mineral resources of the States and Territories west of the Rocky Mountains for 1866, pp. 41–42, 1867.
[63] Raymond, R. W., Statistics of mines and mining in the States and Territories west of the Rocky Mountains for 1870, p. 30, 1872.
[64] Eng. and Min. Jour.-Press, vol. 118, p. 744, 1924.

of the vein and the dip of the slates is well shown at the face of the south drift on the 150-foot level, where it amounts to 30°. The extent of the displacement along the reverse-fault fissure occupied by the vein can not even be estimated at the present time, because no distinctive beds are shown in the mine workings that are accessible.

The ore shoot, as seen on the 150-foot level, is north of the shaft; it is 600 feet long and 2½ feet thick. The vein locally expands to 7 or 8 feet in thickness, but only 2 or 3 feet on the footwall side is mined. That another shoot occurs north of the shoot now being mined is indicated by old surface workings.

The ore averages $7 a ton. It is markedly ribboned throughout, and the ribboning is parallel to the dip of the vein. The ribboning is not a criterion of quality, however, as south of the shaft the vein, although nicely ribboned, proved valueless. In places there is a thick footwall gouge. A single crosscut into the hanging wall of the vein shows that the slates adjacent to the vein are stringered with quartz. Pyrite is the dominant sulphide in the ore, and galena, sphalerite, and tetrahedrite also occur. A notable content of pyrite is the only visible feature of the ore shoot that distinguishes it from the rest of the vein.

INDEX

	Page
Acknowledgments for aid	4
Aegirine-augite, description of	34-35
Albite, description of	35
Albitite, nature and occurrence of	21-22
Albitization of greenstones	16-17, 43
of wall rocks	42, 43-45
Allanite, description of	35
Alteration of wall rocks by the ore-forming solutions	41-45
Amador County, gold mining in	5, 6
Amador Queen No. 2 mine, description of	70
Amphibolite, alteration of	43-44
composition of	11-12
structure of	12
Amphibolite schist associated with the Calaveras formation, age of	12
Ankerite, description of	35-36
Ankerite metasomes in slate, plates showing	40, 41
Antigorite, features of	36
App mine, description of	78-79
Argonaut mine, description of	66-68
Arsenopyrite, description of	36
Augite melaphyre, analysis of	43
description of	15
Basalts, analyses of	16
Bastite, features of	36
"Bedrock series," nature and divisions of	8-9
Bull vein, ore bodies associated with	73, 74-77
Bunker Hill mine, description of	55-56
Calaveras formation, nature and age of	10-11
Calcite, scarcity of	36
Carson Hill, mines on, geology of	73
mines on, history of	72-73
ore bodies in	73-77
Central Eureka mine, description of	59-62
Chalcopyrite, scarcity of	36
Chilena mine, description of	78
Chlorite, features of	36-37
Chlorite schist, alteration of	44
Climate of the belt	1-2
Clio mine, description of	82-83
Conglomerate, alteration of	43
Cost of quartz mining	7
Country rock, mineralized, nature and origin of	33-34
Dolomite. See Ankerite.	
Dutch-Sweeney mine, description of	78-79
Eagle Shawmut mine, description of	79-82
Eureka mine, description of	59
reopening of	7
Fairbanks, H. W., examination of gold belt by	3
Fairchild, J. G., analysis by	43
Faults, cause and extent of	45-46
Field work, record of	3-4
Ford mine, description of	70-71
Fremont mine, description of	52-54
Gabbro, features of	20
Galena, high gold indicated by	37
Galena Slide mine, description of	48-49
Geothermal gradient, measurement of	22-23
Gold, deposits of, features of	23-24
deposits of, origin of	45-48
mode of occurrence of	37
output of	7-8
Gold Cliff mine, description of	71-72
Gold-quartz vein, first discovery of	4
Gouge, nature and occurrence of	26
Granodiorite, description of	21
"Graphitic ore," nature of	39
Gray ore, nature and origin of	33-34

	Page
Graywacke, alteration of	43
occurrence of, in Mariposa slate	13
Greenstone, alteration of	43
inclusions of, in quartz, plate showing	24
mineralized, nature and occurrence of	33
origin of	33-34
Greenstones associated with Mariposa slate, augite basalts in	15-16
contemporaneous intrusions in	17-18
general features of	14-15
keratophyric rocks in	16-17
Guildford mine, description of	49
Gypsum, features of	37
Hillebrand, W. F., analysis by	43
History of the region	4-7
Hornblendite, features of	20-21
Howe, A. S., acknowledgment to	4
Hutchinson, E. C., acknowledgment to	4
Igneous rocks, general relations of	18
Inclusions of greenstone and black slate in quartz, plate showing	24
Kennedy mine, description of	63-66
Keystone mine, description of	58-59
Loring, W. J., acknowledgment to	4
McCraney, O., acknowledgment to	4
Maclaren, Malcolm, quoted	30
Magnetite, occurrence of	37-38
Mammoth mine, description of	69-70
Map geologic, of adit level in Eagle Shawmut mine	80
geologic, of area around the Plymouth mine	56
of the Mother Lode belt	In pocket.
showing location of Mother Lode belt and intrusive rocks of the Sierra Nevada	8
Mariposa estate, gold mining on	4, 5
history and general features of	83-84
mines on	84-85
Mariposa slate, age of	14
alteration of	41-42
conglomerate in	13-14
general features of	12-13
graywacke in	13
petrographic nature of	13
Mariposite, description of	38-39
Melones mine, stereogram of ore shoots in, plate showing	8
See also Carson Hill, mines on.	
Metadiorite, composition and age of	18-19
Metasomes in slate, plates showing	40, 41
Milligan, W. E., analysis by	37
Minerals of the Mother Lode	34-41
Mines, plan of treatment of	48
Mining, history of	4-7
Molybdenite, occurrence of	39
Moore mine, description of	68-69
Morgan mine, stereogram of ore shoots in, plate showing	80
See also Carson Hill, mines on.	
Mother Lode belt, from Carson Hill, plate showing	1
from the Central Eureka mine, plate showing	1
Nolan, Thomas B., Chilena mine	78
Norwegian mine	77
Norwegian mine, description of	77
Nuggets, weight of	37
Ore shoot in Central Eureka mine, plate showing	56
Ore shoots, distribution of gold in	27-28
origin of, early ideas on	29-31
hypothesis for	31-32
influence of wall rocks on	31
situation of	28-29
size and form of	26-27

	Page
Original Amador mine, description of	56-68
Outline of the report	VII-VIII
Physical features of the belt	1
Petzite, description of	39
Plymouth, section across greenstone belt west of	15
Plymouth mine, development of	50
geology of	51-52
history of	49-50
Plymouth mine, output of	50-51
reopening of	6-7
Prehnite, occurrence of	39
Princeton mine, description of	84-85
Pyrrhotite, features of	39
Pyrite, description of	39
metasomatic, in slate, plate showing	41
Quartz, description of	39-40
outcrop of, at Eagle Shawmut mine, plate showing	25
Quartz mining, beginning of	5
Quartz ore, composite, plate showing	25
Quartz vein, typical, gold in and structural features of, plate showing	28
Quartz veins, features of	24-25
Quartzite, alteration of	43
Queen Specimen mine, description of	84
Ransome, F. L., examination of gold belt by	3
Raymond, R. W., quoted	29
Ribbon quartz, plate showing	24
Ribbon structure, indication from	27
Riebeckite, features of	40

	Page
Royal mine, description of	72
Rutile, features of	40
Schists, mineralized, nature and origin of	34
Sericite, description of	40-41
Serpentine, alteration of	44
occurrence and age of	19-20
Sierra Nevada, geology of lower western slope of	8-10
Situation and towns of the Mother Lode belt	1
Smith, W. W., analysis by	36
Sphalerite, features of	41
Solutions, ore-forming, origin of	48
South Eureka mine, description of	62-63
Stanton, T. W., fossils determined by	14
Stokes, H. N., analysis by	22
Storms, W. H., quoted	30
Studies, earlier	3
Swelling ground, cause of	26
Talc, features of	41
Taylor, Bayard, quoted	4
Tellurides, occurrence of	39
Temperature, underground, measurement of	22-23
Tetrahedrite, occurrence of	41
Treasure mine, description of	54-55
Tyson, P. T., examination of gold belt by	4
Union mine, description of	71
Vegetation of the belt	2-3
Veins, filling of	25-26
Water, volume of, in the mines	24-25

○

A. MOTHER LODE BELT, AS SEEN FROM THE CENTRAL EUREKA MINE

The town of Sutter Creek lies amid the trees

B. MOTHER LODE BELT, AS SEEN FROM SUMMIT OF CARSON HILL

Huge white quartz vein of Mother Lode system traverses flank and summit of dark hill in middle ground

NORTH

Vein

Fillon Group

East vein

Shawmut vein

Middle vein

East vein

EXPLANATION

Mariposa slate, with inter-bedded sandstone or graywacke

Augite porphyry

Sheared pyroxenite-chlorite schist, including ferritized rock and quartz

Chlorite-zoisite schist

Calaveras slate

Siliceous slate and quartzite

Stopes

GEOLOGIC MAP OF THE ADIT LEVEL, EAGLE SHAWMUT MINE

STEREOGRAM OF ORE SHOOTS OF MORGAN AND MELONES MINES

1, Ball vein; 2, footwall ore body; 3, hanging-wall ore body; 4, flat vein ore bodies

EXPLANATION

Alluvium

Mariposa slate

Graywacke lenses in the Mariposa slate

Augite melaphyre breccia and tuff lenses in the Mariposa slate

Amphibolite schist

Diabase porphyry dikes

Fault. U upthrow D downthrow

Veins

Mine dump

Strike of vertical schistosity

Strike and dip of schistosity

Strike and dip

NORTH

GEOLOGIC MAP OF THE AREA SURROUNDING THE PLYMOUTH MINE

0   500   1000 Feet

incline of 50°, and the Gover shaft, an incline 1,500 feet in length, which is 1,430 feet north of the Fremont; the workings are connected underground.

thick belt of augitic greenstone tuff and breccia. The greenstone belt, as shown underground, contains some intercalated lenses of slate. All rocks dip 80° E.

FIGURE 8.—Ore shoots of Plymouth mine

The Gover is the older; the Fremont shaft was begun in 1900.

The Fremont shaft is at the surface in Mariposa slate, 50 feet west of the contact of the slate with a

Two veins—a footwall vein, known at the mine as the Mother Lode vein, and a hanging-wall vein—determine the development and exploration policy. The Mother Lode vein is so termed because it contains

A. ANKERITE METASOMES IN SLATE, SURROUNDED BY FRINGES OF CLEAR ANKERITE

B. METASOMATIC PYRITE IN SLATE, SURROUNDED BY FEATHER QUARTZ

A. ANKERITE METASOME IN SLATE, ORDINARY LIGHT

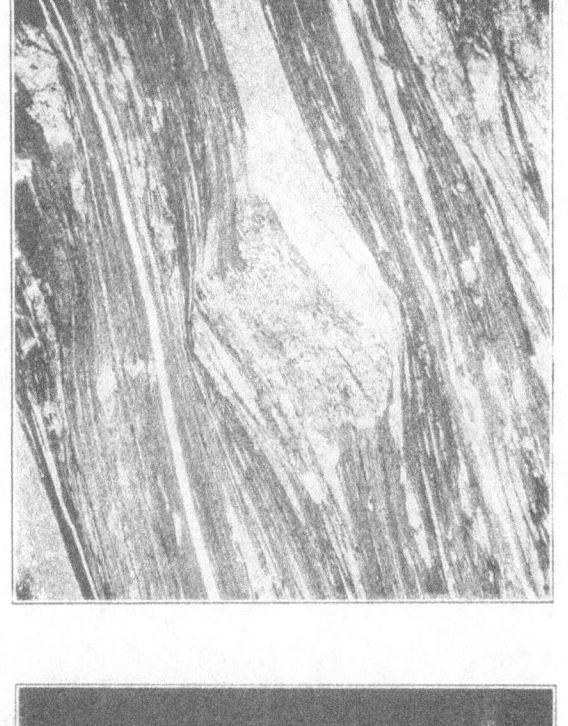

C. METASOMATIC ANKERITE IN SLATE, SHERMAN MINE

B. SAME AS A, BUT IN POLARIZED LIGHT

W.

E.

3,655' LEVEL

3695' 0'-3' (S) 20 c., 3'-6' (S) $1.20
6' @ 70 c.

3704' 0'-2' (S) 20 c.
2' @ 20 c.

3711' 0'-4' (S) 20 c., 4'-6' (S) $1.00
6' @ 45 c.

3716' 0'-4' (S) 80 c., 4'-5' (S) 20 c., 5'-5½' (Q) $39.00
5½' @ $4.15

3723' 0'-2' (S) 20 c., 2'-4' (Q) Tr., 4'-5' (Q) $2.80
5' @ 65 c.

3727' 0'-2' (Q) 20 c., 2'-2½' (Q) $32.00
2½' @ $6.55

3730' 0'-1' (S) 20 c., 1'-2' (Q) $28.40, 2'-3' (G) $3.20
3' @ $10.60

3736' 0'-1' (S) 20 c., 1'-3' (Q) $4.00
3' @ $2.70

3742' 0'-2½' (Q) $5.00
2½' @ $5.00

3746' 0'-2' (S) $8.80, 2'-4' (Q) $15.60
4' @ $12.20

3750' 0'-3' (S) $7.60, 3'-3½' (Q) $96.20, 3½'-5' (Q) $2.80
5' @ $14

3754' 0'-2' (S) $1.20, 2'-2½' (Q) $29.60, 2½'-4' (S) $3.00
4' @ $5.40

3758' 0'-2' (S) $1.60, 2'-2½' (Q) $25.60, 2½'-4' (S) $4.80
4' @ $5.80

3762' 0'-5' (S) $1.60
5' @ $1.60

3772' 0'-1' (Q) - $4.40, 1'-2½' (Q) $45.80, 2½'-4' (Q) $8.80
4' @ $21.60

3776' 0'-2' (Q) $35.20, 2'-4' (Q) $21.40, 4'-5' (G) $5.20
5' @ $22.60

3780' 0'-1' (S) $3.00, 1'-4½' (Q) $36.40, 4½'-5½' (G) $5.80
5½' @ $24.80

3785' 0'-1½' (S) $7.00, 1½'-4½' (Q) $23.20
4½' @ $17.80

3790' 0'-1½' (S) $6.00, 1½'-4' (Q) $18.20
4' @ $13.60

3795' 0'-1½' (S) 80 c., 1½'-3' (Q) $13.60, 3'-4' (G) $4.80
4' @ $7.00

3800' 0'-1½' (S) 60 c.
1½' @ 60 c.

3804' 0'-1½' (S) 40 c.
1½' @ 40 c.

3809' 0'-3' (S) 40 c., 3'-4' (G) $3.80
4' @ $1.20

3811' 0'-3' (Q) $12.85
3' @ $12.85

3817' 0'-3' (S) $12.40, 3'-4½' (Q) $4.00, 4½'-6' (Q) $9.80, 6'-7' (Q) $60.00
7' @ $18.35

3821' 0'-3' (S) $13.80, 3'-6' (Q) $10.80, 6'-7' (Q) $18.40
7' @ $13.00

3825' 0'-1½' (S) $74.80, 1½'-5½' (Q) $24.80, 5½'-6½' (Q) $35.80
6½' @ $38.10

3829' 0'-3' (S) $6.00, 3'-4½' (Q) $10.40, 4½'-5½' (Q) $6.80
5½' @ $7.35

3832' 0'-1½' (S) $18.20, 1½'-5½' (Q) $15.00, 5½'-8' (Q) $11.80
8' @ $14.60

3837' 0'-1½' (Q) $38.00, 1½'-3½' (Q) $12.60, 3½'-7½' (Q) $11.60
7½' @ $17.20

3841' 0'-3½' (Q) $27.60, 3½'-6½' (Q) $8.00
6½' @ $18.55

3844' 0'-1' (Q) $43.00, 1'-3' (Q) $48.40, 3'-5½' (Q) $10.60, 5½'-7' (Q) $18.00
7' @ $27.65

3848' 0'-1' (S) $4.00, 1'-6' (Q) $9.00, 6'-7½' (Q) $12.80
7½' @ $9.25

3852' 0'-3½' (Q) $22.80, 3½'-7½' (Q) $5.00
7½' @ $13.35

3,850' LEVEL

EXPLANATION

Quartz

Gouge

Stringer lode

Mariposa slate

S   Slate
Q   Quartz
G   Gouge

0   10   50 FEET

DISTRIBUTION OF GOLD AND STRUCTURAL FEATURES OF TYPICAL QUARTZ VEIN INCLOSED IN SLATE, PLYMOUTH MINE

A. COMPOSITE QUARTZ ORE FROM 4,200-FOOT LEVEL, ARGONAUT MINE

B. PROMINENT QUARTZ OUTCROP AT EAGLE SHAWMUT MINE

A. RIBBON QUARTZ, PLYMOUTH MINE

B. INCLUSIONS OF GREENSTONE AND BLACK SLATE IN QUARTZ, ORIGINAL
AMADOR MINE

LOCATION OF MOTHER LODE BELT AND OTHER MINING DISTRICTS WITH REFERENCE TO THE INTRUSIVE ROCKS OF THE SIERRA NEVADA

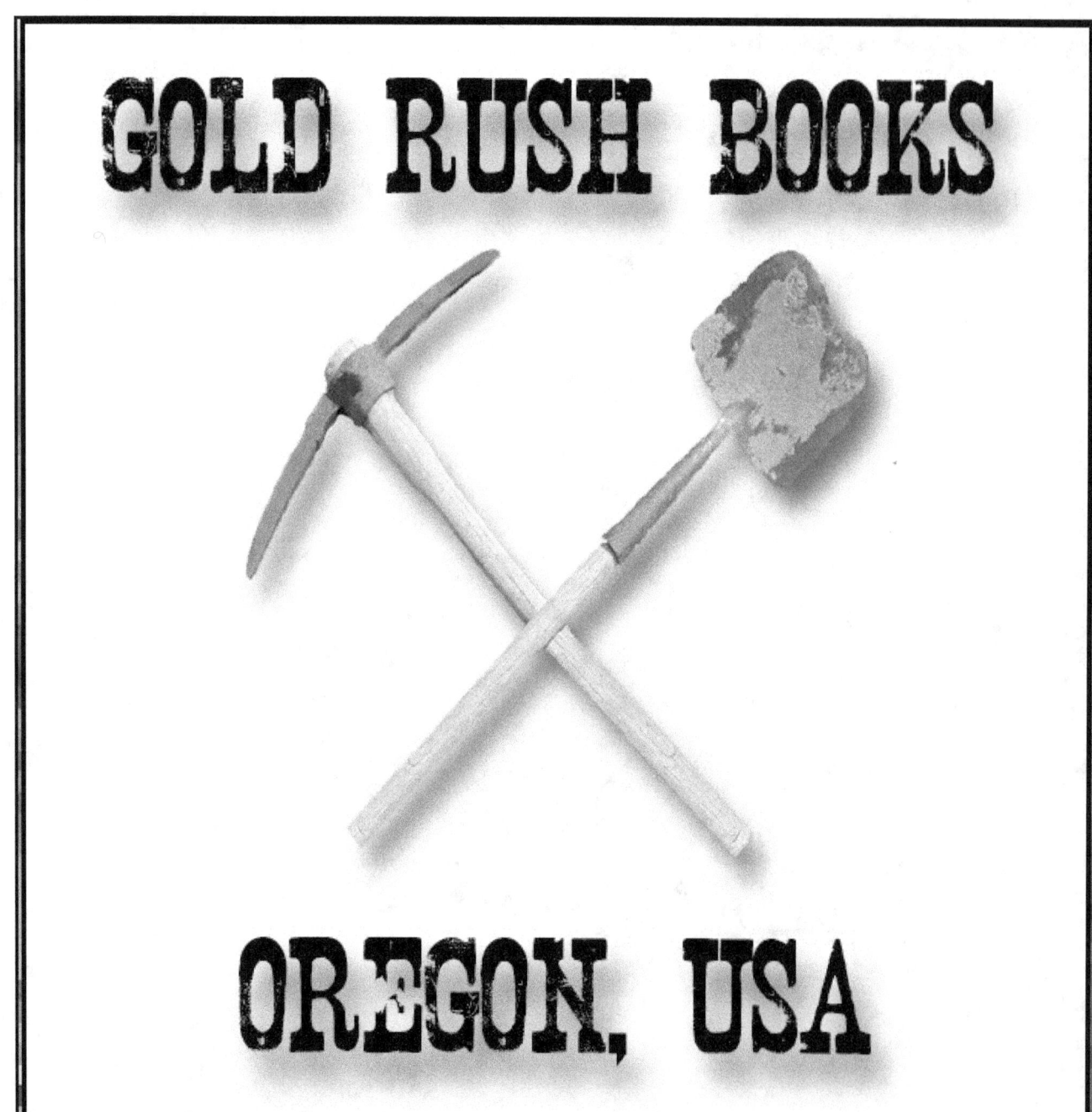

GOLD RUSH BOOKS

OREGON, USA

www.GoldMiningBooks.com

Books On Mining

Visit: www.goldminingbooks.com to order your copies or ask your favorite book seller to offer them.

Mining Books by Kerby Jackson

<u>Gold Dust: Stories From Oregon's Mining Years</u> - Oregon mining historian and prospector, Kerby Jackson, brings you a treasure trove of seventeen stories on Southern Oregon's rich history of gold prospecting, the prospectors and their discoveries, and the breathtaking areas they settled in and made homes. 5" X 8", 98 ppgs. Retail Price: $11.99

<u>The Golden Trail: More Stories From Oregon's Mining Years</u> - In his follow-up to "Gold Dust: Stories of Oregon's Mining Years", this time around, Jackson brings us twelve tales from Oregon's Gold Rush, including the story about the first gold strike on Canyon Creek in Grant County, about the old timers who found gold by the pail full at the Victor Mine near Galice, how Iradel Bray discovered a rich ledge of gold on the Coquille River during the height of the Rogue River War, a tale of two elderly miners on the hunt for a lost mine in the Cascade Mountains, details about the discovery of the famous Armstrong Nugget and others. 5" X 8", 70 ppgs. Retail Price: $10.99

Oregon Mining Books

<u>Geology and Mineral Resources of Josephine County, Oregon</u> - Unavailable since the 1970's, this important publication was originally compiled by the Oregon Department of Geology and Mineral Industries and includes important details on the economic geology and mineral resources of this important mining area in South Western Oregon. Included are notes on the history, geology and development of important mines, as well as insights into the mining of gold, copper, nickel, limestone, chromium and other minerals found in large quantities in Josephine County, Oregon. 8.5" X 11", 54 ppgs. Retail Price: $9.99

<u>Mines and Prospects of the Mount Reuben Mining District</u> - Unavailable since 1947, this important publication was originally compiled by geologist Elton Youngberg of the Oregon Department of Geology and Mineral Industries and includes detailed descriptions, histories and the geology of the Mount Reuben Mining District in Josephine County, Oregon. Included are notes on the history, geology, development and assay statistics, as well as underground maps of all the major mines and prospects in the vicinity of this much neglected mining district. 8.5" X 11", 48 ppgs. Retail Price: $9.99

<u>The Granite Mining District</u> - Notes on the history, geology and development of important mines in the well known Granite Mining District which is located in Grant County, Oregon. Some of the mines discussed include the Ajax, Blue Ribbon, Buffalo, Continental, Cougar-Independence, Magnolia, New York, Standard and the Tillicum. Also included are many rare maps pertaining to the mines in the area. 8.5" X 11", 48 ppgs. Retail Price: $9.99

<u>Ore Deposits of the Takilma and Waldo Mining Districts of Josephine County, Oregon</u> - The Waldo and Takilma mining districts are most notable for the fact that the earliest large scale mining of placer gold and copper in Oregon took place in these two areas. Included are details about some of the earliest large gold mines in the state such as the Llano de Oro, High Gravel, Cameron, Platerica, Deep Gravel and others, as well as copper mines such as the famous Queen of Bronze mine, the Waldo, Lily and Cowboy mines. This volume also includes six maps and 20 original illustrations. 8.5" X 11", 74 ppgs. Retail Price: $9.99

<u>Metal Mines of Douglas, Coos and Curry Counties, Oregon</u> - Oregon mining historian Kerby Jackson introduces us to a classic work on Oregon's mining history in this important re-issue of Bulletin 14C Volume 1, otherwise known as the Douglas, Coos & Curry Counties, Oregon Metal Mines Handbook. Unavailable since 1940, this important publication was originally compiled by the Oregon Department of Geology and Mineral Industries includes detailed descriptions, histories and the geology of over 250 metallic mineral mines and prospects in this rugged area of South West Oregon. 8.5" X 11", 158 ppgs. Retail Price: $19.99

Metal Mines of Jackson County, Oregon - Unavailable since 1943, this important publication was originally compiled by the Oregon Department of Geology and Mineral Industries includes detailed descriptions, histories and the geology of over 450 metallic mineral mines and prospects in Jackson County, Oregon. Included are such famous gold mining areas as Gold Hill, Jacksonville, Sterling and the Upper Applegate. **8.5" X 11", 220 ppgs. Retail Price: $24.99**

Metal Mines of Josephine County, Oregon - Oregon mining historian Kerby Jackson introduces us to a classic work on Oregon's mining history in this important re-issue of Bulletin 14C, otherwise known as the Josephine County, Oregon Metal Mines Handbook. Unavailable since 1952, this important publication was originally compiled by the Oregon Department of Geology and Mineral Industries includes detailed descriptions, histories and the geology of over 500 metallic mineral mines and prospects in Josephine County, Oregon. **8.5" X 11", 250 ppgs. Retail Price: $24.99**

Metal Mines of North East Oregon - Oregon mining historian Kerby Jackson introduces us to a classic work on Oregon's mining history in this important re-issue of Bulletin 14A and 14B, otherwise known as the North East Oregon Metal Mines Handbook. Unavailable since 1941, this important publication was originally compiled by the Oregon Department of Geology and Mineral Industries and includes detailed descriptions, histories and the geology of over 750 metallic mineral mines and prospects in North Eastern Oregon. **8.5" X 11", 310 ppgs. Retail Price: $29.99**

Metal Mines of North West Oregon - Oregon mining historian Kerby Jackson introduces us to a classic work on Oregon's mining history in this important re-issue of Bulletin 14D, otherwise known as the North West Oregon Metal Mines Handbook. Unavailable since 1951, this important publication was originally compiled by the Oregon Department of Geology and Mineral Industries and includes detailed descriptions, histories and the geology of over 250 metallic mineral mines and prospects in North Western Oregon. **8.5" X 11", 182 ppgs. Retail Price: $19.99**

Mines and Prospects of Oregon - Mining historian Kerby Jackson introduces us to a classic mining work by the Oregon Bureau of Mines in this important re-issue of The Handbook of Mines and Prospects of Oregon. Unavailable since 1916, this publication includes important insights into hundreds of gold, silver, copper, coal, limestone and other mines that operated in the State of Oregon around the turn of the 19th Century. Included are not only geological details on early mines throughout Oregon, but also insights into their history, production, locations and in some cases, also included are rare maps of their underground workings. **8.5" X 11", 314 ppgs. Retail Price: $24.99**

Lode Gold of the Klamath Mountains of Northern California and South West Oregon
(See California Mining Books)

Mineral Resources of South West Oregon - Unavailable since 1914, this publication includes important insights into dozens of mines that once operated in South West Oregon, including the famous gold fields of Josephine and Jackson Counties, as well as the Coal Mines of Coos County. Included are not only geological details on early mines throughout South West Oregon, but also insights into their history, production and locations. **8.5" X 11", 154 ppgs. Retail Price: $11.99**

Chromite Mining in The Klamath Mountains of California and Oregon
(See California Mining Books)

Southern Oregon Mineral Wealth - Unavailable since 1904, this rare publication provides a unique snapshot into the mines that were operating in the area at the time. Included are not only geological details on early mines throughout South West Oregon, but also insights into their history, production and locations. Some of the mining areas include Grave Creek, Greenback, Wolf Creek, Jump Off Joe Creek, Granite Hill, Galice, Mount Reuben, Gold Hill, Galls Creek, Kane Creek, Sardine Creek, Birdseye Creek, Evans Creek, Foots Creek, Jacksonville, Ashland, the Applegate River, Waldo, Kerby and the Illinois River, Althouse and Sucker Creek, as well as insights into local copper mining and other topics. **8.5" X 11", 64 ppgs. Retail Price: $8.99**

Geology and Ore Deposits of the Takilma and Waldo Mining Districts - Unavailable since the 1933, this publication was originally compiled by the United States Geological Survey and includes details on gold and copper mining in the Takilma and Waldo Districts of Josephine County, Oregon. The Waldo and Takilma mining districts are most notable for the fact that the earliest large scale mining of placer gold and copper in Oregon took place in these two areas. Included in this report are details about some of the earliest large gold mines in the state such as the Llano de Oro, High Gravel, Cameron, Platerica, Deep Gravel and others, as well as copper mines such as the famous Queen of Bronze mine, the Waldo, Lily and Cowboy mines. In addition to geological examinations, insights are also provided into the production, day to day operations and early histories of these mines, as well as calculations of known mineral reserves in the area. This volume also includes six maps and 20 original illustrations. **8.5" X 11", 74 ppgs. Retail Price: $9.99**

Gold Mines of Oregon - Oregon mining historian Kerby Jackson introduces us to a classic work on Oregon's mining history in this important re-issue of Bulletin 61, otherwise known as "Gold and Silver In Oregon". Unavailable since 1968, this important publication was originally compiled by geologists Howard C. Brooks and Len Ramp of the Oregon Department of Geology and Mineral Industries and includes detailed descriptions, histories and the geology of over 450 gold mines Oregon. Included are notes on the history, geology and gold production statistics of all the major mining areas in Oregon including the Klamath Mountains, the Blue Mountains and the North Cascades. While gold is where you find it, as every miner knows, the path to success is to prospect for gold where it was previously found. **8.5" X 11", 344 ppgs. Retail Price: $24.99**

Mines and Mineral Resources of Curry County Oregon - Originally published in 1916, this important publication on Oregon Mining has not been available for nearly a century. Included are rare insights into the history, production and locations of dozens of gold mines in Curry County, Oregon, as well as detailed information on important Oregon mining districts in that area such as those at Agness, Bald Face Creek, Mule Creek, Boulder Creek, China Diggings, Collier Creek, Elk River, Gold Beach, Rock Creek, Sixes River and elsewhere. Particular attention is especially paid to the famous beach gold deposits of this portion of the Oregon Coast. **8.5" X 11", 140 ppgs. Retail Price: $11.99**

Chromite Mining in South West Oregon - Originally published in 1961, this important publication on Oregon Mining has not been available for nearly a century. Included are rare insights into the history, production and locations of nearly 300 chromite mines in South Western Oregon. **8.5" X 11", 184 ppgs. Retail Price: $14.99**

Mineral Resources of Douglas County Oregon - Originally published in 1972, this important publication on Oregon Mining has not been available for nearly forty years. Included are rare insights into the geology, history, production and locations of numerous gold mines and other mining properties in Douglas County, Oregon. **8.5" X 11", 124 ppgs. Retail Price: $11.99**

Mineral Resources of Coos County Oregon - Originally published in 1972, this important publication on Oregon Mining has not been available for nearly forty years. Included are rare insights into the geology, history, production and locations of numerous gold mines and other mining properties in Coos County, Oregon. **8.5" X 11", 100 ppgs. Retail Price: $11.99**

Mineral Resources of Lane County Oregon - Originally published in 1938, this important publication on Oregon Mining has not been available for nearly seventy five years. Included are extremely rare insights into the geology and mines of Lane County, Oregon, in particular in the Bohemia, Blue River, Oakridge, Black Butte and Winberry Mining Districts. **8.5" X 11", 82 ppgs. Retail Price: $9.99**

Mineral Resources of the Upper Chetco River of Oregon: Including the Kalmiopsis Wilderness - Originally published in 1975, this important publication on Oregon Mining has not been available for nearly forty years. Withdrawn under the 1872 Mining Act since 1984, real insight into the minerals resources and mines of the Upper Chetco River has long been unavailable due to the remoteness of the area. Despite this, the decades of battle between property owners and environmental extremists over the last private mining inholding in the area has continued to pique the interest of those interested in mining and other forms of natural resource use. Gold mining began in the area in the 1850's and has a rich history in this geographic area, even if the facts surrounding it are little known. Included are twenty two rare photographs, as well as insights into the Becca and Morning Mine, the Emmly Mine (also known as Emily Camp), the Frazier Mine, the Golden Dream or Higgins Mine, Hustis Mine, Peck Mine and others. **8.5" X 11", 64 ppgs. Retail Price: $8.99**

Gold Dredging in Oregon - Originally published in 1939, this important publication on Oregon Mining has not been available for nearly seventy five years. Included are extremely rare insights into the history and day to day operations of the dragline and bucketline gold dredges that once worked the placer gold fields of South West and North East Oregon in decades gone by. Also included are details into the areas that were worked by gold dredges in Josephine, Jackson, Baker and Grant counties, as well as the economic factors that impacted this mining method. This volume also offers a unique look into the values of river bottom land in relation to both farming and mining, in how farm lands were mined, re-soiled and reclamated after the dredges worked them. Featured are hard to find maps of the gold dredge fields, as well as rare photographs from a bygone era. **8.5" X 11", 86 ppgs. Retail Price: $8.99**

Quick Silver Mining in Oregon - Originally published in 1963, this important publication on Oregon Mining has not been available for over fifty years. This publication includes details into the history and production of Elemental Mercury or Quicksilver in the State of Oregon. **8.5" X 11", 238 ppgs. Retail Price: $15.99**

Mines of the Greenhorn Mining District of Grant County Oregon - Originally published in 1948, this important publication on Oregon Mining has not been available for over sixty five years. In this publication are rare insights into the mines of the famous Greenhorn Mining District of Grant County, Oregon, especially the famous Morning Mine. Also included are details on the Tempest, Tiger, Bi-Metallic, Windsor, Psyche, Big Johnny, Snow Creek, Banzette and Paramount Mines, as well as prospects in the vicinities in the famous mining areas of Mormon Basin, Vinegar Basin and Desolation Creek. Included are hard to find mine maps and dozens of rare photographs from the bygone era of Grant County's rich mining history. **8.5" X 11", 72 ppgs. Retail Price: $9.99**

Geology of the Wallowa Mountains of Oregon: Part I (Volume 1) - Originally published in 1938, this important publication on Oregon Mining has not been available for nearly seventy five years. Included are details on the geology of this unique portion of North Eastern Oregon. This is the first part of a two book series on the area. Accompanying the text are rare photographs and historic maps.**8.5" X 11", 92 ppgs. Retail Price: $9.99**

Geology of the Wallowa Mountains of Oregon: Part II (Volume 2) - Originally published in 1938, this important publication on Oregon Mining has not been available for nearly seventy five years. Included are details on the geology of this unique portion of North Eastern Oregon. This is the first part of a two book series on the area. Accompanying the text are rare photographs and historic maps.**8.5" X 11", 94 ppgs. Retail Price: $9.99**

Field Identification of Minerals For Oregon Prospectors - Originally published in 1940, this important publication on Oregon Mining has not been available for nearly seventy five years. Included in this volume is an easy system for testing and identifying a wide range of minerals that might be found by prospectors, geologists and rockhounds in the State of Oregon, as well as in other locales. Topics include how to put together your own field testing kit and how to conduct rudimentary tests in the field. This volume is written in a clear and concise way to make it useful even for beginners. **8.5" X 11", 158 ppgs. Retail Price: $14.99**

The Bohemia Mining District of Oregon - Originally published in 1900, this important publication on Oregon Mining has not been available for over a century. Included in this volume are important insights into the famous Bohemia Mining District of Oregon, including the histories and locations of important gold mines in the area such as the Ophir Mine, Clarence, Acturas, Peek-a-boo, White Swan, Combination Mine, the Musick Mine, The California, White Ghost, The Mystery, Wall Street, Vesuvius, Story, Lizzie Bullock, Delta, Elsie Dora, Golden Slipper, Broadway, Champion Mine, Knott, Noonday, Helena, White Wings, Riverside and others. Also included are notes on the nearby Blue River Mining District. **8.5" X 11", 58 ppgs. Retail Price: $9.99**

The Gold Fields of Eastern Oregon - Unavailable since 1900, this publication was originally compiled by the Baker City Chamber of Commerce Offering important insights into the gold mining history of Eastern Oregon, "The Gold Fields of Eastern Oregon" sheds a rare light on many of the gold mines that were operating at the turn of the 19th Century in Baker County and Grant County in North Eastern Oregon. Some of the areas featured include the Cable Cove District, Baisely-Elhorn, Granite, Red Boy, Bonanza, Susanville, Sparta, Virtue, Vaughn, Sumpter, Burnt River, Rye Valley and other mining districts. Included is basic information on not only many gold mines that are well known to those interested in Eastern Oregon mining history, but also many mines and prospects which have been mostly lost to the passage of time. Accompanying are numerous rare photos **8.5" X 11", 78 ppgs. Retail Price: $10.99**

Gold Mining in Eastern Oregon - Originally published in 1938, this important publication on Oregon Mining has not been available for over a century. Included in this volume are important insights into the famous mining districts of Eastern Oregon during the late 1930's. Particular attention is given to those gold mines with milling and concentrating facilities in the Greenhorn, Red Boy, Alamo, Bonanza, Granite, Cable Cove, Cracker Creek, Virtue, Keating, Medical Springs, Sanger, Sparta, Chicken Creek, Mormon Basin, Connor Creek, Cornucopia and the Bull Run Mining Districts. Some of the mines featured include the Ben Harrison, North Pole-Columbia, Highland Maxwell, Baisley-Elkhorn, White Swan, Balm Creek, Twin Baby, Gem of Sparta, New Deal, Gleason, Gifford-Johnson, Cornucopia, Record, Bull Run, Orion and others. Of particular interest are the mill flow sheets and descriptions of milling operations of these mines. **8.5" X 11", 68 ppgs. Retail Price: $8.99**

The Gold Belt of the Blue Mountains of Oregon - Originally published in 1901, this important publication on Oregon Mining has not been available for over a century. Included in this volume are rare insights into the gold deposits of the Blue Mountains of North East Oregon, including the history of their early discovery and early production. Extensive details are offered on this important mining area's mineralogy and economic geology, as well as insights into nearby gold placers, silver deposits and copper deposits. Featured are the Elkhorn and Rock Creek mining districts, the Pocahontas district, Auburn and Minersville districts, Sumpter and Cracker Creek, Cable Cove, the Camp Carson district, Granite, Alamo, Greenhorn, Robinsonville, the Upper Burnt River Valley and Bonanza districts, Susanville, Quartzburg, Canyon Creek, Virtue, the Copper Butte district, the North Powder River, Sparta, Eagle Creek, Cornucopia, Pine Creek, Lower Powder River, the Upper Snake River Canyon, Rye Valley, Lower Burnt River Valley, Mormon Basin, the Malheur and Clarks Creek districts, Sutton Creek and others. Of particular interest are important details on numerous gold mines and prospects in these mining districts, including their locations, histories, geology and other important information, as well as information on silver, copper and fire opal deposits. **8.5" X 11", 250 ppgs. Retail Price: $24.99**

Mining in the Cascades Range of Oregon - Originally published in 1938, this important publication on Oregon Mining has not been available for over seventy five years. Included in this volume are rare insights into the gold mines and other types of metal mines in the Cascades Mountain Range of Oregon. Some of the important mining areas covered include the famous Bohemia Mining District, the North Santiam Mining District, Quartzville Mining District, Blue River Mining District, Fall Creek Mining District, Oakridge District, Zinc District, Buzzard-Al Sarena District, Grand Cove, Climax District and Barron Mining District. Of particular interest are important details on over 100 mines and prospects in these mining districts, including their locations, histories, geology and other important information. 8.5" X 11", 170 ppgs. Retail Price: $14.99

Beach Gold Placers of the Oregon Coast - Originally published in 1934, this important publication on Oregon Mining has not been available for over 80 years. Included in this volume are rare insights into the beach gold deposits of the State of Oregon, including their locations, occurance, composition and geology. Of particular interest is information on placer platinum in Oregon's rich beach deposits. Also included are the locations and other information on some famous Oregon beach mines, including the Pioneer, Eagle, Chickamin, Iowa and beach placer mines north of the mouth of the Rogue River. 8.5" X 11", 60 ppgs. Retail Price: $8.99

Mineralogical Composition of the Sands of the Oregon Coast: From Coos Bay to the Columbia - Published in 1945, he text features hard to find information on the composition of the gold bearing black sands of the South West Oregon Coast, offering a unique insight to prospectors in search of Oregon's legendary beach gold. 104 ppgs, $9.99

Manganese Mining in Oregon - First released in 1942 and now out of print, this special reprint edition of "Manganese in Oregon" was originally published by the Oregon Department of Geology and Mineral Industries. The text features hard to find information on the mining of Manganese in Oregon, including details and maps of Oregon manganese mines and prospects. 108 ppgs, 9.99

Medford Oregon As A Mining Center - Written in 1912, this hard to find publication includes valuable insights into the mining history of South West Oregon. This small book contains interesting information on the gold, copper and mining industry in Southern Oregon as it existed just prior to World War One, shedding light on some of the important mines in the area. Included are rare photographs and vintage advertising of the day. 80 ppgs, 9.99

Mineral Resources of Curry County Oregon - First released in 1977 and now out of print, this special reprint edition of "Geology, Mineral Resources and Rock Materials of Curry County, Oregon" was originally published in cooperation of Curry County, Oregon and the Oregon Department of Geology and Mineral Industries. The text features hard to find information on not only the mining of gold and other metals in Curry County, but also aggregate mining in the area. 102 ppgs, 11.99

Origin of the Gold Bearing Black Sands of the Coast of South West Oregon - First released in 1943 and now out of print, this special reprint edition of "The Origin of the Black Sands of the South West Oregon Coast" was originally published by the Oregon Department of Geology and Mineral Industries. The text features hard to find information on the origin of the gold bearing black sands of the South West Oregon Coast, offering a unique insight to prospectors in search of Oregon's legendary beach gold. 52 ppgs, 8.99

South West Oregon Mining - Leading mining historian Kerby Jackson introduces us to six classic small mining publications on the Gold Mining Industry in Southern Oregon. This small book consists of a compilation of USGS J.S. Diller's "Mines of the Riddles Quadrangle", "The Rogue River Valley Coal Fields" and "Mineral Resources of the Grants Pass Quadrangle", the Grants Pass Commercial Club's rare publication "Mining in Josephine County, Oregon" and the USGS publication "The Distribution of Placer Gold in the Sixes River, South West Oregon". Also included is F.W. Libbey's legendary article on the Southern Oregon Mining Industry, "Lest We Forget", which appeared in the publication of the Oregon State Department of Geology and Mineral Industries in the early 1960's. This compilation offers a unique perspective on mining in South West Oregon and includes considerable information on mines in Josephine, Jackson and Coos Counties. 142 ppgs, 14.99

Geology and Mineral Resources of the Gasquet Quadrangle of California-Oregon - First published in 1953, it has been unavailable for over a century and sheds important light on the geological features and mineral resources of this portion of Northern California and Southern Oregon. 80 ppgs, 9.99

Idaho Mining Books

Gold in Idaho - Unavailable since the 1940's, this publication was originally compiled by the Idaho Bureau of Mines and includes details on gold mining in Idaho. Included is not only raw data on gold production in Idaho, but also valuable insight into where gold may be found in Idaho, as well as practical information on the gold bearing rocks and other geological features that will assist those looking for placer and lode gold in the State of Idaho. This volume also includes thirteen gold maps that greatly enhance the practical usability of the information contained in this small book detailing where to find gold in Idaho. **8.5" X 11", 72 ppgs. Retail Price: $9.99**

Geology of the Couer D'Alene Mining District of Idaho - Unavailable since 1961, this publication was originally compiled by the Idaho Bureau of Mines and Geology and includes details on the mining of gold, silver and other minerals in the famous Coeur D'Alene Mining District in Northern Idaho. Included are details on the early history of the Coeur D'Alene Mining District, local tectonic settings, ore deposit features, information on the mineral belts of the Osburn Fault, as well as detailed information on the famous Bunker Hill Mine, the Dayrock Mine, Galena Mine, Lucky Friday Mine and the infamous Sunshine Mine. This volume also includes sixteen hard to find maps. **8.5" X 11", 70 ppgs. Retail Price: $9.99**

The Gold Camps and Silver Cities of Idaho - Originally published in 1963, this important publication on Idaho Mining has not been available for nearly fifty years. Included are rare insights into the history of Idaho's Gold Rush, as well as the mad craze for silver in the Idaho panhandle. Documented in fine detail are the early mining excitements at Boise Basin, at South Boise, in the Owyhees, at Deadwood, Long Valley, Stanley Basin and Robinson Bar, at Atlanta, on the famous Boise River, Volcano, Little Smokey, Banner, Boise Ridge, Hailey, Leesburg, Lemhi, Pearl, at South Mountain, Shoup and Ulysses, Yellow Jacket and Loon Creek. The story follows with the appearance of Chinese miners at the new mining camps on the Snake River, Black Pine, Yankee Fork, Bay Horse, Clayton, Heath, Seven Devils, Gibbonsville, Vienna and Sawtooth City. Also included are special sections on the Idaho Lead and Silver mines of the late 1800's, as well as the mining discoveries of the early 1900's that paved the way for Idaho's modern mining and mineral industry. Lavishly illustrated with rare historic photos, this volume provides a one of a kind documentary into Idaho's mining history that is sure to be enjoyed by not only modern miners and prospectors who still scour the hills in search of nature's treasures, but also those enjoy history and tromping through overgrown ghost towns and long abandoned mining camps. **8.5" X 11", 186 ppgs. Retail Price: $14.99**

Ore Deposits and Mining in North Western Custer County Idaho - Unavailable since 1913, this important publication was originally published by the Us Department of the Interior and has been unavailable for a century. Included are fine details on the geology, geography, gold placers and gold and silver bearing quartz veins of the mining region of North West Custer County, Idaho. Of particular interest is a rare look at the mines and prospects of the region, including those such as the Ramshorn Mine, SkyLark, Riverview, Excelsior, Beardsley, Pacific, Hoosier, Silver Brick, Forest Rose and dozens of others in the Bay Horse Mining District. Also covered are the mines of the Yankee Fork District such as the Lucky Boy, Badger, Black, Enterprise, Charles Dickens, Morrison, Golden Sunbeam, Montana, Golden Gate and others, as well as those in the Loon Mining District. **8.5" X 11", 126 ppgs. Retail Price: $12.99**

Gold Rush To Idaho - Unavailable since 1963, this important publication was originally published by the Idaho Bureau of Mines and has been unavailable for 50 years. "Gold Rush To Idaho" revisits the earliest years of the discovery of gold in Idaho Territory and introduces us to the conditions that the pioneer gold seekers met when they blazed a trail through the wilderness of Idaho's mountains and discovered the precious yellow metal at Oro Fino and Pierce. Subsequent rushes followed at places like Elk City, Newsome, Clearwater Station, Florence, Warrens and elsewhere. Of particular interest is a rare look at the hardships that the first miners in Idaho met with during their day to day existences and their attempts to bring law and order to their mining camps. **8.5" X 11", 88 ppgs. Retail Price: $9.99**

The Geology and Mines of Northern Idaho and North Western Montana - Unavailable since 1909, this important publication was originally published by the Us Department of the Interior and has been unavailable for a century. Included are fine details on the geology and geography of the mining regions of Northern Idaho and North Western Montana. Of particular interest is a rare look at the mines and prospects of the region, including those in the Pine Creek Mining District, Lake Pend Oreille district, Troy Mining District, Sylvanite District, Cabinet Mining District, Prospect Mining District and the Missoula Valley. Some of the mines featured include the Iron Mountain, Silver Butte, Snowshoe, Grouse Mountain Mine and others. **8.5" X 11", 142 ppgs. Retail Price: $12.99**

Mining in the Alturas Quadrangle of Blaine County Idaho - Unavailable since 1922, this important publication was originally published by the Idaho Bureau of Mines and has been unavailable for ninety years. Topics include the geology, rock formations and the formation of ore deposits in this important mining area of Idaho. Of particular focus is information on the local geology, quartz veins and ore deposits of this portion of Idaho. Included are hard to find details, including the descriptions and locations of numerous gold and silver mines in the area including the Silver King, Pilgrim, Columbia, Lone Jack, Sunbeam, Pride of the West, Lucky Boy, Scotia, Atlanta, Beaver-Bidwell and others mines and prospects. **8.5" X 11", 56 ppgs. Retail Price: $8.99**

Mining in Lemhi County Idaho - Originally published in 1913, this important book on Idaho Mining has not been available to miners for over a century. Included are rare insights into hundreds of gold, silver, copper and other mines in this famous Idaho mining area. Details include the locations, geology, history, production and other facts of the mines of this region, not only gold and silver hardrock mines, but also gold placer mines, lead-silver deposits, copper mines, cobalt-nickel deposits, tungsten and tin mines . It is lavishly illustrated with hard to find photos of the period and rare mining maps. Some of the vicinities featured include the Nicholia Mining District, Spring Mountain District, Texas District, Blue Wing District, Junction District, McDevitt District, Pratt Creek, Eldorado District, Kirtley Creek, Carmen Creek, Gibbonsville, Indian Creek, Mineral Hill District, Mackinaw, Eureka District, Blackbird District, YellowJacket District, Gravel Range District, Junction District, Parker Mountain and other mining districts. **8.5" X 11", 226 ppgs. Retail Price: $19.99**

Mining in Shoshone County Idaho - First published in 1923, it has been unavailable for over a century and sheds important light on the mining history of Shoshone County, Idaho. Some of the topics include the history of mining in Shoshone County, a look at the local geology and ore characteristics of lead-silver deposits, zinc deposits, copper, antimony, gold and other minerals. Also included are insights into the history, production, characteristics and locations of numerous mines in the area. 198 ppgs, 15.99

Utah Mining Books

Fluorite in Utah - Unavailable since 1954, this publication was originally compiled by the USGS, State of Utah and U.S. Atomic Energy Commission and details the mining of fluorspar, also known as fluorite in the State of Utah. Included are details on the geology and history of fluorspar (fluorite) mining in Utah, including details on where this unique gem mineral may be found in the State of Utah. **8.5" X 11", 60 ppgs. Retail Price: $8.99**

The Gold Hill Mining District of Utah - First published in 1935, it has been unavailable since those days and sheds important light on the mines, history and geology of Utah's Gold Hill Mining District. Included are rare insights into this important mining area, including the locations, histories and details of numerous mines. This volume is well illustrated with geological diagrams, as well as hard to find maps of some of the most important mines in this district. 202 ppgs., 19.99

The Mines, Miners and Minerals of Utah - First published in 1896, it has been unavailable since those days and sheds important light on the early mines and miners of Pioneer Utah, as well as the minerals which they won from the earth by laborious hard physical labor and sheer determination. Included are rare insights into the early mining history of Utah, as well details on hundreds of gold, silver and copper mines. 376 ppgs., 24.99

California Mining Books

The Tertiary Gravels of the Sierra Nevada of California - Mining historian Kerby Jackson introduces us to a classic mining work by Waldemar Lindgren in this important re-issue of The Tertiary Gravels of the Sierra Nevada of California. Unavailable since 1911, this publication includes details on the gold bearing ancient river channels of the famous Sierra Nevada region of California. **8.5" X 11", 282 ppgs. Retail Price: $19.99**

The Mother Lode Mining Region of California - Unavailable since 1900, this publication includes details on the gold mines of California's famous Mother Lode gold mining area. Included are details on the geology, history and important gold mines of the region, as well as insights into historic mining methods, mine timbering, mining machinery, mining bell signals and other details on how these mines operated. Also included are insights into the gold mines of the California Mother Lode that were in operation during the first sixty years of California's mining history. **8.5" X 11", 176 ppgs. Retail Price: $14.99**

Lode Gold of the Klamath Mountains of Northern California and South West Oregon - Unavailable since 1971, this publication was originally compiled by Preston E. Hotz and includes details on the lode mining districts of Oregon and California's Klamath Mountains. Included are details on the geology, history and important lode mines of the French Gulch, Deadwood, Whiskeytown, Shasta, Redding, Muletown, South Fork, Old Diggings, Dog Creek (Delta), Bully Choop (Indian Creek), Harrison Gulch, Hayfork, Minersville, Trinity Center, Canyon Creek, East Fork, New River, Denny, Liberty (Black Bear), Cecilville, Callahan, Yreka, Fort Jones and Happy Camp mining districts in California, as well as the Ashland, Rogue River, Applegate, Illinois River, Takilma, Greenback, Galice, Silver Peak, Myrtle Creek and Mule Creek districts of South Western Oregon. Also included are insights into the mineralization and other characteristics of this important mining region. **8.5" X 11", 100 ppgs. Retail Price: $10.99**

Mines and Mineral Resources of Shasta County, Siskiyou County, Trinity County: California - Unavailable since 1915, this publication was originally compiled by the California State Mining Bureau and includes details on the gold mines of this area of Northern California. Also included are insights into the mineralization and other characteristics of this important mining region, as well as the location of historic gold mines. **8.5" X 11", 204 ppgs. Retail Price: $19.99**

Geology of the Yreka Quadrangle, Siskiyou County, California - Unavailable since 1977, this publication was originally compiled by Preston E. Hotz and includes details on the geology of the Yreka Quadrangle of Siskiyou County, California. Also included are insights into the mineralization and other characteristics of this important mining region. **8.5" X 11", 78 ppgs. Retail Price: $7.99**

Mines of San Diego and Imperial Counties, California - Originally published in 1914, this important publication on California Mining has not been available for a century. This publication includes important information on the early gold mines of San Diego and Imperial County, which were some of the first gold fields mined in California by early Spanish and Mexican miners before the 49ers came on the scene. Included are not only details on early mining methods in the area, production statistics and geological information, but also the location of the early gold mines that helped make California "The Golden State". Also included are details on the mining of other minerals such as silver, lead, zinc, manganese, tungsten, vanadium, asbestos, barite, borax, cement, clay, dolomite, fluospar, gem stones, graphite, marble, salines, petroleum, stronium, talc and others. **8.5" X 11", 116 ppgs. Retail Price: $12.99**

Mines of Sierra County, California - Unavailable since 1920, this publication was originally compiled by the California State Mining Bureau and includes details on the gold mines of Sierra County, California. Also included are insights into the mineralization and other characteristics of this important mining region, as well as the location of historic gold mines. **8.5" X 11", 156 ppgs. Retail Price: $19.99**

Mines of Plumas County, California - Unavailable since 1918, this publication was originally compiled by the California State Mining Bureau and includes details on the gold mines of Plumas County, California. Also included are insights into the mineralization and other characteristics of this important mining region, as well as the location of historic gold mines. **8.5" X 11", 200 ppgs. Retail Price: $19.99**

Mines of El Dorado, Placer, Sacramento and Yuba Counties, California - Originally published in 1917, this important publication on California Mining has not been available for nearly a century. This publication includes important information on the early gold mines of El Dorado County, Placer County, Sacramento County and Yuba County, which were some of the first gold fields mined by the Forty-Niners during the California Gold Rush. Included are not only details on early mining methods in the area, production statistics and geological information, but also the location of the early gold mines that helped make California "The Golden State". Also included are insights into the early mining of chrome, copper and other minerals in this important mining area. **8.5" X 11", 204 ppgs. Retail Price: $19.99**

Mines of Los Angeles, Orange and Riverside Counties, California - Originally published in 1917, this important publication on California Mining has not been available for nearly a century. This publication includes important information on the early gold mines of Los Angeles County, Orange County and Riverside County, which were some of the first gold fields mined in California by early Spanish and Mexican miners before the 49ers came on the scene. Included are not only details on early mining methods in the area, production statistics and geological information, but also the location of the early gold mines that helped make California "The Golden State". **8.5" X 11", 146 ppgs. Retail Price: $12.99**

Mines of San Bernadino and Tulare Counties, California - Originally published in 1917, this important publication on California Mining has not been available for nearly a century. This publication includes important information on the early gold mines of San Bernadino and Tulare County, which were some of the first gold fields mined in California by early Spanish and Mexican miners before the 49ers came on the scene. Included are not only details on early mining methods in the area, production statistics and geological information, but also the location of the early gold mines that helped make California "The Golden State". Also included are details on the mining of other minerals such as copper, iron, lead, zinc, manganese, tungsten, vanadium, asbestos, barite, borax, cement, clay, dolomite, fluospar, gem stones, graphite, marble, salines, petroleum, stronium, talc and others. **8.5" X 11", 200 ppgs. Retail Price: $19.99**

Chromite Mining in The Klamath Mountains of California and Oregon - Unavailable since 1919, this publication was originally compiled by J.S. Diller of the United States Department of Geological Survey and includes details on the chromite mines of this area of Northern California and Southern Oregon. Also included are insights into the mineralization and other characteristics of this important mining region, as well as the location of historic mines. Also included are insights into chromite mining in Eastern Oregon and Montana. **8.5" X 11", 98 ppgs. Retail Price: $9.99**

Mines and Mining in Amador, Calaveras and Tuolumne Counties, California - Unavailable since 1915, this publication was originally compiled by William Tucker and includes details on the mines and mineral resources of this important California mining area. Included are details on the geology, history and important gold mines of the region, as well as insights into other local mineral resources such as asbestos, clay, copper, talc, limestone and others. Also included are insights into the mineralization and other characteristics of this important portion of California's Mother Lode mining region. **8.5" X 11", 198 ppgs. Retail Price: $14.99**

The Cerro Gordo Mining District of Inyo County California - Unavailable since 1963, this publication was originally compiled by the United States Department of Interior. Included are insights into the mineralization and other characteristics of this important mining region of Southern California. Topics include the mining of gold and silver in this important mining district in Inyo County, California, including details on the history, production and locations of the Cerro Gordo Mine, the Morning Star Mine, Estelle Tunnel, Charles Lease Tunnel, Ignacio, Hart, Crosscut Tunnel, Sunset, Upper Newtown, Newtown, Ella, Perseverance, Newsboy, Belmont and other silver and gold mines in the Cerro Gordo Mining District. This volume also includes important insights into the fossil record, geologic formations, faults and other aspects of economic geology in this California mining district. 8.5" X 11", 104 ppgs. Retail Price: $10.99

Mining in Butte, Lassen, Modoc, Sutter and Tehama Counties of California - Unavailable since 1917, this publication was originally compiled by the United States Department of Interior. Included are insights into the mineralization and other characteristics of this important mining region of California. Topics include the mining of asbestos, chromite, gold, diamonds and manganese in Butte County, the mining of gold and copper in the Hayden Hill and Diamond Mountain mining districts of Lassen County, the mining of coal, salt, copper and gold in the High Grade and Winters mining districts of Modoc County, gold mining in Sutter County and the mining of gold, chromite, manganese and copper in Tehama County. This volume also includes the production records and locations of numerous mines in this important mining region. 8.5" X 11", 114 ppgs. Retail Price: $11.99

Mines of Trinity County California - Originally published in 1965, this important publication on California Mining has not been available for nearly fifty years. This publication includes important information on mines and mining in Trinity County, California, as well insights into the mineralization and geology of this important mining area in Northern California. Included are extensive details on hardrock and placer gold mines and prospects, including charts showing the locations of these historic mines.. 8.5" X 11", 144 ppgs. Retail Price: $12.99

Mines of Kern County California - Originally published in 1962, this important publication on California Mining has not been available for nearly fifty years. This publication includes important information on mines and mining in Kern County, California, as well insights into the mineralization and geology of this important mining area in California. Included are extensive details on hardrock and placer gold mines and prospects, including charts showing the locations of these historic mines. 8.5" X 11", 398 ppgs. Retail Price: $24.99

Mines of Calaveras County California - Originally published in 1962, this important publication on California Mining has not been available for nearly fifty years. This publication includes important information on mines and mining in Calaveras County, California, as well insights into the mineralization and geology of this important mining area in Northern California. Included are extensive details on hardrock and placer gold mines and prospects, including charts showing the locations of these historic mines. 8.5" X 11", 236 ppgs. Retail Price: $19.99

Lode Gold Mining in Grass Valley California - Unavailable since 1940, this publication was originally compiled by the United States Department of Interior. Included are insights into the gold mineralization and other characteristics of this important mining region of Nevada County, California. This volume also includes important insights into the geologic formations, faults and other aspects of economic geology in this California mining district. Of particular interest are the fine details on many hardrock gold mines in the area, including their locations, histories, development and mineralization. Some of the mines featured include the Gold Hill Mine, Massachusetts Hill, Boundary, Peabody, Golden Center, North Star, Omaha, Lone Jack, Homeward Bound, Hartery, Wisconsin, Allison Ranch, Phoenix, Kate Hayes, W.Y.O.D., Empire, Rich Hill, Daisy Hill, Orleans, Sultana, Centennial, Conlin, Ben Franklin, Crown Point and many others. 8.5" X 11", 148 ppgs. Retail Price: $12.99

Lode Mining in the Alleghany District of Sierra County California - Unavailable since 1913, this publication was originally compiled by the United States Department of Interior. Included are insights into the mineralization and other characteristics of this important mining region of Sierra County. Included are details on the history, production and locations of numerous hardrock gold mines in this famous California area, including the Tightner Mine, Minnie D., Osceola, Eldorado, Twenty One, Sherman, Kenton, Oriental, Rainbow, Plumbago, Irelan, Gold Canyon, North Fork, Federal, Kate Hardy and others. This volume also includes important insights into the fossil record, geologic formations, faults and other aspects of economic geology in this California mining district. 8.5" X 11", 48 ppgs. Retail Price: $7.99

Six Months In The Gold Mines During The California Gold Rush - Unavailable since 1850, this important work is a first hand account of one "49'ers" personal experience during the great California Gold Rush, shedding important light on one of the most exciting periods in the history of not only California, but also the world. Compiled from journals written between 1847 and 1849 by E. Gould Buffum, a native of New York, "Six Months In The Gold Mines During The California Gold Rush" offers a rare look into the day to day lives of the people who came to California to work in her gold mines when the state was still a great frontier. 8.5" X 11", 290 ppgs. Retail Price: $19.99

Quartz Mines of the Grass Valley Mining District of California - Unavailable since 1867, this important publication has not been available since those days. This rare publication offers a short dissertation on the early hardrock mines in this important mining district in the California Mother Lode region between the 1850's and 1860's. Also included are hard to find details on the mineralization and locations of these mines, as well as how they were operated in those day. **8.5" X 11", 44 ppgs. Retail Price: $8.99**

Gold Rush on the Feather River - First published in 1924, this short publication by G.C. Mansfield sheds important light on the early history of gold mining on the Feather River. Included are rare insights into the first decade of gold mining and the early mining camps of the Feather River during the 1850's. 64 ppgs., 9.99

The Bodie Mining District of California - First published in 1986, it has been unavailable since those days and sheds important light on this famous mining area. Included are the history, characteristics and locations of numerous old mines around the ghost town of Bodie. 64 ppgs, 8.99

Geology and Mineral Resources of the Gasquet Quadrangle of California-Oregon - First published in 1953, it has been unavailable for over a century and sheds important light on the geological features and mineral resources of this portion of Northern California and Southern Oregon. 80 ppgs, 9.99

Alaska Mining Books

Ore Deposits of the Willow Creek Mining District, Alaska - Unavailable since 1954, this hard to find publication includes valuable insights into the Willow Creek Mining District near Hatcher Pass in Alaska. The publication includes insights into the history, geology and locations of the well known mines in the area, including the Gold Cord, Independence, Fern, Mabel, Lonesome, Snowbird, Schroff-O'Neil, High Grade, Marion Twin, Thorpe, Webfoot, Kelly-Willow, Lane, Holland and others. **8.5" X 11", 96 ppgs. Retail Price: $9.99**

The Juneau Gold Belt of Alaska - Unavailable since 1906, this hard to find publication includes valuable insights into the gold mines around Juneau, Alaska. The publication includes important details into the history, geology and locations of the well known gold mines and prospects in the area, including those around Windham Bay, Holkham Bay, Port Snettisham, on Grindstone and Rhine Creeks, Gold Creek, Douglas Island, Salmon Creek, Lemon Creek, Nugget Creek, from the Mendenhall River to Berners Bay, McGinnis Creek, Montana Creek, Peterson Creek, Windfall Creek, the Eagle River, Yankee Basin, Yankee Curve, Kowee Creek and elsewhere. Not only are gold placer mines included, but also hardrock gold mines. **8.5" X 11", 224 ppgs. Retail Price: $19.99**

Mining in the Jumbo Basin of Alaska - Unavailable since 1953, this hard to find publication includes valuable insights into the mines and geology of the Jumbo Basin. The publication includes important details into the history, geology and locations of the well known gold mines and prospects in the famous Jumbo Basin Mining Region of Alaska. 72 ppgs, 9.99

The Rampart Placer Gold Region of Alaska - Unavailable since 1906, this hard to find publication includes valuable insights into the placer gold mines of the Rampart Mining Region. The publication includes important details into the history, geology and locations of the well known gold mines and prospects in the famous Rampart Mining Region of Alaska. 78 ppgs, 10.99

Arizona Mining Books

Mines and Mining in Northern Yuma County Arizona - Originally published in 1911, this important publication on Arizona Mining has not been available for over a hundred years. Included are rare insights into the gold, silver, copper and quicksilver mines of Yuma County, Arizona together with hard to find maps and photographs. Some of the mines and mining districts featured include the Planet Copper Mine, Mineral Hill, the Clara Consolidated Mine, Viati Mine, Copper Basin prospect, Bowman Mine, Quartz King, Billy Mack, Carnation, the Wardwell and Osbourne, Valensuella Copper, the Mariquita, Colonial Mine, the French American, the New York-Plomosa, Guadalupe, Lead Camp, Mudersbach Copper Camp, Yellow Bird, the Arizona Northern (Salome Strike), Bonanza (Harqua Hala), Golden Eagle, Hercules, Socorro and others. **8.5" X 11", 144 ppgs. Retail Price: $11.99**

The Aravaipa and Stanley Mining Districts of Graham County Arizona - Originally published in 1925, this important publication on Arizona Mining has not been available for nearly ninety years. Included are rare insights into the gold and silver mines of these two important mining districts, together with hard to find maps. **8.5" X 11", 140 ppgs. Retail Price: $11.99**

Gold in the Gold Basin and Lost Basin Mining Districts of Mohave County, Arizona - This volume contains rare insights into the geology and gold mineralization of the Gold Basin and Lost Basin Mining Districts of Mohave County, Arizona that will be of benefit to miners and prospectors. Also included is a significant body of information on the gold mines and prospects of this portion of Arizona. This volume is lavishly illustrated with rare photos and mining maps. **8.5" X 11", 188 ppgs. Retail Price: $19.99**

Mines of the Jerome and Bradshaw Mountains of Arizona - This important publication on Arizona Mining has not been available for ninety years. This volume contains rare insights into the geology and ore deposits of the Jerome and Bradshaw Mountains of Arizona that will be of benefit to miners and prospectors who work those areas. Included is a significant body of information on the mines and prospects of the Verde, Black Hills, Cherry Creek, Prescott, Walker, Groom Creek, Hassayampa, Bigbug, Turkey Creek, Agua Fria, Black Canyon, Peck, Tiger, Pine Grove, Bradshaw, Tintop, Humbug and Castle Creek Mining Districts. This volume is lavishly illustrated with rare photos and mining maps. **8.5" X 11", 218 ppgs. Retail Price: $19.99**

The Ajo Mining District of Pima County Arizona - This important publication on Arizona Mining has not been available for nearly seventy years. This volume contains rare insights into the geology and mineralization of the Ajo Mining District in Pima County, Arizona and in particular the famous New Cornelia Mine. **8.5" X 11", 126 ppgs. Retail Price: $11.99**

Mining in the Santa Rita and Patagonia Mountains of Arizona - Originally published in 1915, this important publication on Arizona Mining has not been available for nearly a century. Included are rare insights into hundreds of gold, silver, copper and other mines in this famous Arizona mining area. Details include the locations, geology, history, production and other facts of the mines of this region. **8.5" X 11", 394 ppgs. Retail Price: $24.99**

Mining in the Bisbee Quadrangle of Arizona - Originally published in 1906, this important publication on Arizona Mining has not been available for nearly a century. Included are rare insights into hundreds of gold, silver, copper and other mines in this famous Arizona mining area. Details include the locations, geology, history, production and other facts of the mines of this important mining region. **8.5" X 11", 188 ppgs. Retail Price: $14.99**

Placer Gold Mining in Arizona - Unavailable since 1922, this hard to find publication includes valuable insights into the placer gold mines of the Arizona. Originally released as "Placer Gold of Arizona", despite its small size, this publication includes important details into the history, geology and locations of the well known placer gold mines and prospects in the State of Arizona. **48 ppgs, 8.99**

Gold and Copper Mining near Payson, Arizona - Written in 1915, this hard to find publication includes valuable insights into the gold and copper mining industry of Arizona. Highlighted here are the gold and copper mines near Payson, Arizona. **68 ppgs, 8.99**

Lode Gold Mining in Arizona - Unavailable since 1934, this hard to find publication, originally released as "Arizona Lode Gold Mines and Gold Mining" includes valuable insights into the gold mining industry of Arizona. Included are valuable insights into over 150 hardrock gold mines in over 30 different mining districts in Arizona. **278 ppgs, 21.99**

Mining in the Dragoon Quadrangle of Cochise County, Arizona - Unavailable since 1964, this hard to find publication includes valuable insights into the mines of the Dragoon Quadrangle Mining Region. The publication includes important details into the history, geology and locations of the well known mines and prospects in this famous mining region of Arizona. **224 ppgs., 19.99**

Directory of Operating Mines in Arizona in 1915 - Unavailable since 1916, this hard to find publication includes valuable insights into the mines of Arizona. This small publication includes a complete list of the mines that were operating in the State of Arizona during 1915 and includes details such as general location, owners and some basic facts about each mining operation. **52 ppgs. 8.99**

Arizona Ore Deposits - Unavailable since 1938, this hard to find publication includes valuable insights into some ore deposits of Arizona. Included are valuable insights into the formation and characteristics of valuable ore deposits in the Jerome, Miami, Inspiration, Clifton, Morenci, Ray, Ajo, Eureka, Tombstone and Magma mining districts. Included are details into some of the major gold, silver and copper mines of these important Arizona mining areas. **160 ppgs, 14.99**

Montana Mining Books

A History of Butte Montana: The World's Greatest Mining Camp - First published in 1900 by H.C. Freeman, this important publication sheds a bright light on one of the most important mining areas in the history of The West. Together with his insights, as well as rare photographs of the periods, Harry Freeman describes Butte and its vicinity from its early beginnings, right up to its flush years when copper flowed from its mines like a river. At the time of publication, Butte, Montana was known worldwide as "The Richest Mining Spot On Earth" and produced not only vast amounts of copper, but also silver, gold and other metals from its mines. Freeman illustrates, with great detail, the most important mines in the vicinity of Butte, providing rare details on their owners, their history and most importantly, how the mines operated and how their treasures were extracted. Of particular interest are the dozens of rare photographs that depict mines such as the famous Anaconda, the Silver Bow, the Smoke House, Moose, Paulin, Buffalo, Little Minah, the Mountain Consolidated, West Greyrock, Cora, the Green Mountain, Diamond, Bell, Parnell, the Neversweat, Nipper, Original and many others. 8.5" X 11", 142 ppgs. Retail Price: $12.99

The Butte Mining District of Montana - This important publication on Montana Mining has not been available for over a century. Included are rare insights into the gold, copper and silver mines of Butte, Montana together with hard to find maps and photographs. Some of the topics include the early history of gold, silver and copper mining in the Butte area, insight into the geology of its mining areas, the local distribution of gold, silver and copper ores, as well their composition and how to identify them. Also included are detailed facts about the mines in the Butte Mining District, including the famous Anaconda Mine, Gagnon, Parrot, Blue Vein, Moscow, Poulin, Stella, Buffalo, Green Mountain, Wake Up Jim, the Diamond-Bell Group, Mountain Consolidated, East Greyrock, West Greyrock, Snowball, Corra, Speculator, Adirondack, Miners Union, the Jessie-Edith May Group, Otisco, Iduna, Colorado, Lizzie, Cambers, Anderson, Hesperus, Preferencia and dozens of others. 8.5" X 11", 298 ppgs. Retail Price: $24.99

Mines of the Helena Mining Region of Montana - This important publication on Montana Mining has not been available for over a century. Included are rare insights into the gold, copper and silver mines of the vicinity of Helena, Montana, including the Marysville Mining District, Elliston Mining District, Rimini Mining District, Helena Mining District, Clancy Mining District, Wickes Mining District, Boulder and Basin Mining Districts and the Elkhorn Mining District. Some of the topics include the early history of gold, silver and copper mining in the Helena area, insight into the geology of its mining areas, the local distribution of gold, silver and copper ores, as well their composition and how to identify them. Also included are detailed facts, history, geology and locations of over one hundred gold, silver and copper mines in the area . 8.5" X 11", 162 ppgs, Retail Price: $14.99

Mines and Geology of the Garnet Range of Montana - This important publication on Montana Mining has not been available for over a century. Included are rare insights into the gold, copper and silver mines of the vicinity of this important mining area of Montana. Some of the topics include the early history of gold, silver and copper mining in the Garnet Mountains, insight into the geology of its mining areas, the local distribution of gold, silver and copper ores, as well their composition and how to identify them. Also included are detailed facts, history, geology and locations of numerous gold, silver and copper mines in the area . 8.5" X 11", 100 ppgs, Retail Price: $11.99

Mines and Geology of the Philipsburg Quadrangle of Montana - This important publication on Montana Mining has not been available for over a century. Included are rare insights into the gold, copper and silver mines of the vicinity of this important mining area of Montana. Some of the topics include the early history of gold, silver and copper mining in the Philipsburg Quadrangle, insight into the geology of its mining areas, the local distribution of gold, silver and copper ores, as well their composition and how to identify them. Also included are detailed facts, history, geology and locations of over one hundred gold, silver and copper mines in the area 8.5" X 11", 290 ppgs, Retail Price: $24.99

Geology of the Marysville Mining District of Montana - Included are rare insights into the mining geology of the Marysville Mining District. Some of the topics include the early history of gold, silver and copper mining in the area, insight into the geology of its mining areas, the local distribution of gold, silver and copper ores, as well their composition and how to identify them. Also included are detailed facts, history, geology and locations of gold, silver and copper mines in the area 8.5" X 11", 198 ppgs, Retail Price: $19.99

The Geology and Mines of Northern Idaho and North Western Montana- See listing under Idaho.

The History of Gold Dredging in Montana - Unavailable since 1916, this important publication was originally published by the Us Bureau of Mines and has been unavailable for a century. A century and more ago, giant dredging machines dug in Montana's rivers and creeks in search of illusive golden riches. First appearing in California in the 1850's, gold dredges finally reached their peak of development in Siberia and New Zealand before becoming popular again in the United States. This book offers a unique historical perspective on the gold dredges that once operated in Montana. This book on Montana mining history is lavishly illustrated with dozens of rare historic photos gold dredges that once operated in Montana, as well as hard to locate plans on how these dredges were designed. 120 ppgs., 11.99

Nevada Mining Books

The Bull Frog Mining District of Nevada - Unavailable since 1910, this publication was originally compiled by the United States Department of Interior. This volume also includes important insights into the geologic formations, faults and other aspects of economic geology in this Nevada mining district. Of particular interest are the fine details on many mines in the area, including their locations, histories, development and mineralization. Some of the mines featured include the National Bank Mine, Providence, Gibraltor, Tramps, Denver, Original Bullfrog, Gold Bar, Mayflower, Homestake-King and other mines and prospects. **8.5" X 11", 152 ppgs, Retail Price: $14.99**

History of the Comstock Lode - Unavailable since 1876, this publication was originally released by John Wiley & Sons. This volume also includes important insights into the famous Comstock Lode of Nevada that represented the first major silver discovery in the United States. During its spectacular run, the Comstock produced over 192 million ounces of silver and 8.2 million ounces of gold. Not only did the Comstock result in one of the largest mining rushes in history and yield immense fortunes for its owners, but it made important contributions to the development of the State of Nevada, as well as neighboring California. Included here are important details on not only the early development and history of the Comstock, but also rare early insight into its mines, ore and its geology.8.5" X 11", 244 ppgs, Retail Price: $19.99

The Pioche Mining District of Nevada - First published in 1932, it has been unavailable for over a century and sheds important light on the mining history of Nevada. Some of the topics include the history of mining in this district, as well as the characteristics of its mineral and ore deposits. Also included are insights into the history, production, characteristics and locations of numerous mines in the area. Some of the mines include the Combined Metals, Pioche, Ely Valley, No. 10, Poorman, Wide Awake, Alps, Prince, Virginia Louise, Half Moon, Abe Lincoln, Fairview, Bristol Silver, National, Vesuvius, Inman, Tempest, Hillside, Jackrabbit, Lucky Star, Fortuna, Mendha, Manhattan, Hamburg, Comet, Lyndon and others. 108 ppgs 10.99

The Yerington Mining District of Nevada - First published in 1932, it has been unavailable for over a century and sheds important light on the mining history of Nevada. Some of the topics include the history of mining in this district, as well as the characteristics of its mineral and ore deposits. Also included are insights into the history, production, characteristics and locations of numerous mines in the area. Some of the mines include the Bluestone, Mason Valley, Malachite, McConnell, Greenwood, Western Nevada, Ludwig, Douglas Hill, Casting Copper, Montana-Yerington, Empire, Jim Beatty, Terry and McFarland, Blue Jay and others. 92 ppgs, 10.99

The Genesis of the Ores of Tonopah Nevada - Unavailable since 1918, this hard to find publication includes valuable insights into the gold mines around Tonopah, Nevada. The publication includes important details into the geology of mines in the Tonopah Mining District of Nevada. 90 ppgs, 10.99

Mining Camps of Elko, Lander and Eureka Counties Nevada - Unavailable since 1910, this hard to find publication includes valuable insights into the mining camps of Elko, Lander and Eureka Counties, Nevada. The publication includes important details into the history of mines and mining in these three Nevada counties. 154 ppgs, 12.99

Ore Deposits of the Bullfrog Quadrangle - Unavailable since 1964 and released as "Geology of Bullfrog Quadrangle and Ore Deposits Related to Bullfrog Hills Caldera, Nye County, Nevada and Inyo County, California". The publication includes important details into the geology of mines in the Bullfrog Quadrangle of Nye County, Nevada and Inyo County, California. 52 ppgs, 9.99

Mining in Eureka County Nevada - Unavailable since 1879, this hard to find publication includes valuable insights into the early mining history off Eureka County, Nevada. The publication includes important details into the early history of the mines of Eureka County, as well as their development, production and how their ores were treated. Also included are details on the 1872 Mining Act, as well as the local rules, regulations and customs of the miners in Eureka County.134 ppgs, 12.99

Colorado Mining Books

Ores of The Leadville Mining District - Unavailable since 1926, this publication was originally compiled by the United States Department of Interior. This volume also includes important insights into the ores and mineralization of the Leadville Mining District in Colorado. Topics include historic ore prospecting methods, local geology, insights into ore veins and stockworks, the local trend and distribution of ore channels, reverse faults, shattered rock above replacement ore bodies, mineral enrichment in oxidized and sulphide zones and more. **8.5" X 11", 66 ppgs, Retail Price: $8.99**

Mining in Colorado - Unavailable since 1926, this publication was originally compiled by the United States Department of Interior. This volume also includes important insights into the mining history of Colorado from its early beginnings in the 1850's right up to the mid 1920's. Not only is Colorado's gold mining heritage included, but also its silver, copper, lead and zinc mining industry. Each mining area is treated separately, detailing the development of Colorado's mines on a county by county basis. **8.5" X 11", 284 ppgs, Retail Price: $19.99**

Gold Mining in Gilpin County Colorado - Unavailable since 1876, this publication was originally compiled by the Register Steam Printing House of Central City, Colorado. A rare glimpse at the gold mining history and early mines of Gilpin County, Colorado from their first discovery in the 1850's up to the "flush years" of the mid 1870's. Of particular interest is the history of the discovery of gold in Gilpin County and details about the men who made those first strikes. Special focus is given to the early gold mines and first mining districts of the area, many of which are not detailed in other books on Colorado's gold mining history. **8.5" X 11", 156 ppgs, Retail Price: $12.99**

Mining in the Gold Brick Mining District of Colorado - Important insights into the history of the Gold Brick Mining District, as well as its local geography and economic geology. Also included are the histories and locations of historic mines in this important Colorado Mining District, including the Cortland, Carter, Raymond, Gold Links, Sacramento, Bassick, Sandy Hook, Chronicle, Grand Prize, Chloride, Granite Mountain, Lucille, Gray Mountain, Hilltop, Maggie Mitchell, Silver Islet, Revenue, Roosevelt, Carbonate King and others. In addition to hardrock mining, are also included are details on gold placer mining in this portion of Colorado. **8.5" X 11", 140 ppgs, Retail Price: $12.99**

Ore Deposits of the London Fault of Colorado - First published in 1941, it has been unavailable since those days and sheds important light on the mines and mineral deposits of the London Fault in Central Colorado's Alma Mining District. This publication sheds important light on the gold veins and lead-silver deposits of the Alma Mining District. Included are geologic details on the **London Mine, American Mine, Havigorst Tunnel, Ophir Mine, Mosher Tunnel, London-Butte Mine, Venture Shaft, Hard-To-Beat Mine, Oliver Twist Tunnel, Sacramento Mine, Mudsill Mine, Sherwood Mine, Wagner, Barcoe Tunnel** and other mines in this important mining region. 110 ppgs., 10.99

The Mines of Colorado - First published in 1867, it has been unavailable since those days and sheds important light on Colorado's early mining history. Written shortly after the events took place, this publication sheds important light on the Pike's Peak Gold Rush, the discovery of gold on Ralston Creek and Dry Creek in the 1850's, as well as details on the first wave of miners into Colorado and their trials and tribulations as they crossed the Great Plains. Also included are details on early discoveries of lode gold in the mountainous regions of Colorado, details on the early mines hardrock and placer mines, and much more. It is a veritable treasure trove on Colorado's early mining history and will be of great importance to anyone who is interested in the mining of gold or other minerals in Colorado, as well as those interested in the history of the state. 478 ppgs., 29.99

The La Plata Mining District of Colorado - Originally titled "Geology and Ore Deposits in the Vicinity of the La Plata District of Colorado" and first published in 1949, it has been unavailable since those days and sheds important light on the mines and mineral deposits of the La Plata Mining District of Colorado. 214 ppgs., 19.99

Washington Mining Books

The Republic Mining District of Washington - Unavailable since 1910, this important publication was originally published by the Washington Geologic Survey and has been unavailable for a century. Topics include the geology, rock formations and the formation of ore deposits in this important mining area of Washington State. Also included are hard to find details on the geology, history and locations of dozens of mines in the area. Some of the mines featured include the New Republic Mine, Ben Hur, Morning Glory, the South Republic Mine, Quilp, Surprise, Black Tail, Lone Pine, San Poil, Mountain Lion, Tom Thumb, Elcaliph and many others. **8.5" X 11", 94 ppgs, Retail Price: $10.99**

The Myers Creek and Nighthawk Mining Districts of Washington - Unavailable since 1911, this important publication was originally published by the Washington Geologic Survey and has been unavailable for a century. Topics include the geology, rock formations and the formation of ore deposits in these important mining areas of Washington State. Also included are hard to find details on the geology, history and locations of dozens of mines in the area. Some of the mines featured include the Grant Mine, Monterey, Nip and Tuck, Myers Creek, Number Nine, Neutral, Rainbow, Aztec, Crystal Butte, Apex, Butcher Boy, Molson, Mad River, Olentangy, Delate, Kelsey, Golden Chariot, Okanogan, Ohio, Forty-Ninth Parallel, Nighthawk, Favorite, Little Chopaka, Summit, Number One, California, Peerless, Caaba, Prize Group, Ruby, Mountain Sheep, Golden Zone, Rich Bar, Similkameen, Kimberly, Triune, Hiawatha, Trinity, Hornsilver, Maquae, Bellevue, Bullfrog, Palmer Lake, Ivanhoe, Copper World and many others. **8.5" X 11", 136 ppgs, Retail Price: $12.99**

The Blewett Mining District of Washington - Unavailable since 1911, this important publication was originally published by the Washington Geologic Survey and has been unavailable for a century. Topics include the geology, rock formations and the formation of ore deposits in this important mining area of Washington State. Also included are hard to find details on the geology, history and locations of dozens of mines in the area. Some of the mines featured include the Washington Meteor, Alta Vista, Pole Pick, Blinn, North Star, Golden Eagle, Tip Top, Wilder, Golden Guinea, Lucky Queen, Blue Bell, Prospect, Homestake, Lone Rock, Johnson, and others. **8.5" X 11", 134 ppgs, Retail Price: $12.99**

Silver Mining In Washington - Unavailable since 1955, this important publication was originally published by the Washington Geologic Survey. Featured are the hard to find locations and details pertaining to Washington's silver mines. **8.5" X 11", 180 ppgs, Retail Price: $15.99**

The Mines of Snohomish County Washington - Unavailable since 1942, this important publication was originally published by the Washington Geologic Survey and has been unavailable for seventy years. Featured are details on a large number of gold, silver, copper, lead and other metallic mineral mines. Included are the locations of each historic mine, along with information on the commodity produced. **8.5" X 11", 98 ppgs, Retail Price: $10.99**

The Mines of Chelan County Washington - Unavailable since 1943, this important publication was originally published by the Washington Geologic Survey and has been unavailable for seventy years. Featured are details on a large number of gold, silver, copper, lead and other metallic mineral mines. Included are the locations of each historic mine, along with information on the commodity. **8.5" X 11", 88 ppgs, Retail Price: $9.99**

Metal Mines of Washington - Unavailable since 1921, this important publication was originally published by the Washington Geologic Survey and has been unavailable for nearly ninety years. Widely considered a masterpiece on the Washington Mining Industry, "Metal Mines of Washington" sheds light on the important details of Washington's early mining years. Featured are details on hundreds of gold, silver, copper, lead and other metallic mineral mines. Included are hard to find details on the mineral resources of this state, as well as the locations of historic mines. Lavishly illustrated with maps and historic photos and complete with a glossary to explain any technical terms found in the text, this is one of the most important works on mining in the State of Washington. No prospector or miner should be without it if they are interested in mining in Washington. **8.5" X 11", 396 ppgs, Retail Price: $24.99**

Gem Stones In Washington - Unavailable since 1949, this important publication was originally published by the Washington Geologic Survey and has been unavailable since first published. Included are details on where to find naturally occurring gem stones in the State of Washington, including quartz crystal, amethyst, smoky quartz, milky quartz, agates, bloodstone, carnelian, chert, flint, jasper, onyx, petrified wood, opal, fire opal, hyalite and others. **8.5" X 11", 54 ppgs, Retail Price: $8.99**

The Covada Mining District of Washington - Unavailable since 1913, this important publication was originally published by the Washington Geologic Survey and has been unavailable for a century. Topics include the geology, rock formations and the formation of ore deposits in this important mining area of Washington State. Also included are hard to find details on the geology, history and locations of dozens of mines in the area. Some of the mines featured include the Admiral, Advance, Algonkian, Big Bug, Big Chief, Big Joker, Black Hawk, Black Tail, Black Thorn, Captain, Cherokee Strip, Colorado, Dan Patch, Dead Shot, Etta, Good Ore, Greasy Run, Great Scott, Idora, IXL, Jay Bird, Kentucky Bell, King Solomon, Laurel, Laura S, Little Jay, Meteor, Neglected, Northern Light, Old Nell, Plymouth Rock, Polaris, Quandary, Reserve, Shoo Fly, Silver Plume, Three Pines, Vernie, White Rose and dozens of others. **8.5" X 11", 114 ppgs, Retail Price: $10.99**

The Index Mining District of Washington - Unavailable since 1912, this important publication was originally published by the Washington Geologic Survey and has been unavailable for a century. Topics include the geology, rock formations and the formation of ore deposits in this important mining area of Washington State. Also included are hard to find details on the geology, history and locations of dozens of mines in the area. Some of the mines featured include the Sunset, Non-Pareil, Ethel Consolidated, Kittaning, Merchant, Homestead, Co-operative, Lost Creek, Uncle Sam, Calumet, Florence-Rae, Bitter Creek, Index Peacock, Gunn Peak, Helena, North Star, Buckeye. Copper Bell, Red Cross and others. **8.5" X 11", 114 ppgs, Retail Price: $11.99**

Mining & Mineral Resources of Stevens County Washington - Unavailable since 1920, this important publication was originally published by the Washington Geologic Survey and has been unavailable for a century. Topics include the geology, rock formations and the formation of ore deposits in these important mining areas of Washington State. Also included are hard to find details on the geology, history and locations of hundreds of mines in the area. **8.5" X 11", 372 ppgs, Retail Price: $24.99**

The Mines and Geology of the Loomis Quadrangle Okanogan County, Washington - Unavailable since 1972, this important publication was originally published by the Washington Geologic Survey and has been unavailable for a century. Topics include the geology, rock formations and the formation of ore deposits in this important mining area of Washington State. Also included are hard to find details on the geology, history and locations of dozens of gold, copper, silver and other mines in the area. **8.5" X 11", 150 ppgs, Retail Price: $12.99**

The Conconully Mining District of Okanogan County Washington - Unavailable since 1973, this important publication was originally published by the Washington Geologic Survey and has been unavailable for a century. Topics include the geology, rock formations and the formation of ore deposits in this important mining area of Washington State, which also includes Salmon Creek, Blue Lake and Galena. Also included are hard to find details on the geology, mining history and locations of dozens of mines in the area. Some of the mines include Arlington, Fourth of July, Sonny Boy, First Thought, Last Chance, War Eagle-Peacock, Wheeler, Mohawk, Lone Star, Woo Loo Moo Loo, Keystone, Hughes, Plant-Callahan, Johnny Boy, Leuena, Gubser, John Arthur, Tough Nut, Homestake, Key and many others **8.5" X 11", 68 ppgs, Retail Price: $8.99**

Wyoming Mining Books

Mining in the Laramie Basin of Wyoming - Unavailable since 1909, this publication was originally compiled by the United States Department of Interior. Also included are insights into the mineralization and other characteristics of this important mining region, especially in regards to coal, limestone, gypsum, bentonite clay, cement, sand, clay and copper. **8.5" X 11", 104 ppgs, Retail Price: $11.99**

New Mexico Mining Books

The Mogollon Mining District of New Mexico - Unavailable since 1927, this important publication was originally published by the US Department of Interior and has been unavailable for 80 years. Topics include the geology, rock formations and the formation of ore deposits in this important mining area in New Mexico. Of particular focus is information on the history and production of the ore deposits in this area, their form and structure, vein filling, their paragenesis, origins and ore shoots, as well as oxidation and supergene enrichment. Also included are hard to find details, including the descriptions and locations of numerous gold, silver and other types of mines, including the Eureka, Pacific, South Alpine, Great Western, Enterprise, Buffalo, Mountain View, Floride, Gold Dust, Last Chance, Deadwood, Confidence, Maud S., Deep Down, Little Fanney, Trilby, Johnson, Alberta, Comet, Golden Eagle, Cooney, Queen, the Iron Crown, Eberle, Clifton, Andrew Jackson mine, Mascot and others. **8.5" X 11", 144 ppgs, Retail Price: $12.99**

The Percha Mining District of Kingston New Mexico - Unavailable since 1883, this important publication was originally published by the Kingston Tribune and has been unavailable for over one hundred and thirty five years. Having been written during the earliest years of gold and silver mining in the Percha Mining District, unlike other books on the subject, this work offers the unique perspective of having actually been written while the early mining history of this area was still being made. In fact, the work was written so early in the development of this area that many of the notable mines in the Percha District were less than a few years old and were still being operated by their original discoverers with the same enthusiasm as when they were first located. Included are hard to find details on the very earliest gold and silver mines of this important mining district near Kingston in Sierra County, New Mexico. **8.5" X 11", 68 ppgs, Retail Price: $9.99**

East Coast Mining Books

The Gold Fields of the Southern Appalachians - Unavailable since 1895, this important publication was originally published by the US Department of Interior and has been unavailable for nearly 120 years. Topics include the geology, rock formations and the formation of ore deposits in this important mining area of the American South. Of particular focus is information on the history and statistics of the ore deposits in this area, their form and structure and veins. Also included are details on the placer gold deposits of the region. The gold fields of the Georgian Belt, Carolinian Belt and the South Mountain Mining District of North Carolina are all treated in descriptive detail. Included are hard to find details, including the descriptions and locations of numerous gold mines in Georgia, North Carolina and elsewhere in the American South. Also included are details on the gold belts of the British Maritime Provinces and the Green Mountains. **8.5" X 11", 104 ppgs, Retail Price: $9.99**

Gold Rush Tales Series

Millions in Siskiyou County Gold - In this first volume of the "Gold Rush Tales" series, leading mining historian and editor Kerby Jackson, introduces us to the story of how millions of dollars worth of gold was discovered in Siskiyou County during the California Gold Rush. Lavishly illustrated with photos from the 19th Century, this hard to find information was first published in 1897 and sheds important light onto the gold rush era in Siskiyou County, California and the experiences of the men who dug for the gold and actually found it. **8.5" X 11", 82 ppgs, Retail Price: $9.99**

The California Rand in the Days of '49 - In this second volume of the "Gold Rush Tales" series, leading mining historian and editor Kerby Jackson, introduces us to four tales from the California Gold Rush. Lavishly illustrated with photos from the 19th Century, this hard to find information was first published in 1890's and includes the stories of "California's Rand", details about Chinese miners, how one early miner named Baker struck it rich and also the story of Alphonzo Bowers, who invented the first hydraulic gold dredge. **8.5" X 11", 54 ppgs, Retail Price: $9.99**

More Mining Books

Prospecting and Developing A Small Mine - Topics covered include the classification of varying ores, how to take a proper ore sample, the proper reduction of ore samples, alluvial sampling, how to understand geology as it is applied to prospecting and mining, prospecting procedures, methods of ore treatment, the application of drilling and blasting in a small mine and other topics that the small scale miner will find of benefit. **8.5" X 11", 112 ppgs, Retail Price: $11.99**

Timbering For Small Underground Mines - Topics covered include the selection of caps and posts, the treatment of mine timbers, how to install mine timbers, repairing damaged timbers, use of drift supports, headboards, squeeze sets, ore chute construction, mine cribbing, square set timbering methods, the use of steel and concrete sets and other topics that the small underground miner will find of benefit. This volume also includes twenty eight illustrations depicting the proper construction of mine timbering and support systems that greatly enhance the practical usability of the information contained in this small book. **8.5" X 11", 88 ppgs. Retail Price: $10.99**

Timbering and Mining - A classic mining publication on Hard Rock Mining by W.H. Storms. Unavailable since 1909, this rare publication provides an in depth look at American methods of underground mine timbering and mining methods. Topics include the selection and preservation of mine timbers, drifting and drift sets, driving in running ground, structural steel in mine workings, timbering drifts in gravel mines, timbering methods for driving shafts, positioning drill holes in shafts, timbering stations at shafts, drainage, mining large ore bodies by means of open cuts or by the "Glory Hole" system, stoping out ore in flat or low lying veins, use of the "Caving System", stoping in swelling ground, how to stope out large ore bodies, Square Set timbering on the Comstock and its modifications by California miners, the construction of ore chutes, stoping ore bodies by use of the "Block System", how to work dangerous ground, information on the "Delprat System" of stoping without mine timbers, construction and use of headframes and much more. This volume provides a reference into not only practical methods of mining and timbering that may be employed in narrow vein mining by small miners today, but also rare insights into how mines were being worked at the turn of the 19th Century. **8.5" X 11", 288 ppgs. Retail Price: $24.99**

A Study of Ore Deposits For The Practical Miner - Mining historian Kerby Jackson introduces us to a classic mining publication on ore deposits by J.P. Wallace. First published in 1908, it has been unavailable for over a century. Included are important insights into the properties of minerals and their identification, on the occurrence and origin of gold, on gold alloys, insights into gold bearing sulfides such as pyrites and arsenopyrites, on gold bearing vanadium, gold and silver tellurides, lead and mercury tellurides, on silver ores, platinum and iridium, mercury ores, copper ores, lead ores, zinc ores, iron ores, chromium ores, manganese ores, nickel ores, tin ores, tungsten ores and others. Also included are facts regarding rock forming minerals, their composition and occurrences, on igneous, sedimentary, metamorphic and intrusive rocks, as well as how they are geologically disturbed by dikes, flows and faults, as well as the effects of these geologic actions and why they are important to the miner. Written specifically with the common miner and prospector in mind, the book will help to unlock the earth's hidden wealth for you and is written in a simple and concise language that anyone can understand. **8.5″ X 11″, 366 ppgs. Retail Price: $24.99**

Mine Drainage - Unavailable since 1896, this rare publication provides an in depth look at American methods of underground mine drainage and mining pump systems. This volume provides a reference into not only practical methods of mining drainage that may be employed in narrow vein mining by small miners today, but also rare insights into how mines were being worked at the turn of the 19th Century. **8.5″ X 11″, 218 ppgs. Retail Price: $24.99**

Fire Assaying Gold, Silver and Lead Ores - Unavailable since 1907, this important publication was originally published by the Mining and Scientific Press and was designed to introduce miners and prospectors of gold, silver and lead to the art of fire assaying. Topics include the fire assaying of ores and products containing gold, silver and lead; the sampling and preparation of ore for an assay; care of the assay office, assay furnaces; crucibles and scorifiers; assay balances; metallic ores; scorification assays; cupelling; parting' crucible assays, the roasting of ores and more. This classic provides a time honored method of assaying put forward in a clear, concise and easy to understand language that will make it a benefit to even beginners. **8.5″ X 11″, 96 ppgs. Retail Price: $11.99**

Methods of Mine Timbering - Originally published in 1896, this important publication on mining engineering has not been available for nearly a century. Included are rare insights into historical methods of timbering structural support that were used in underground metal mines during the California that still have a practical application for the small scale hardrock miner of today. **8.5″ X 11″, 94 ppgs. Retail Price: $10.99**

The Enrichment of Copper Sulfide Ores - First published in 1913, it has been unavailable for over a century. Topics include the definition and types of ore enrichment, the oxidation of copper ores, the precipitation of metallic sulfides. Also included are the results of dozens of lab experiments pertaining to the enrichment of sulfide ores that will be of interest to the practical hard rock mine operator in his efforts to release the metallic bounty from his mine's ore. **8.5″ X 11″, 92 ppgs. Retail Price: $9.99**

A Study of Magmatic Sulfide Ores - Unavailable since 1914, this rare publication provides an in depth look at magmatic sulfide ores. Some of the topics included are the definition and classification of magmatic ores, descriptions of some magmatic sulfide ore deposits known at the time of publication including copper and nickel bearing pyrrohitic ore bodies, chalcopyrite-bornite deposits, pyritic deposits, magnetite-ileminite deposits, chromite deposits and magmatic iron ore deposits. Also included are details on how to recognize these types of ore deposits while prospecting for valuable hardrock minerals. **8.5″ X 11″, 138 ppgs. Retail Price: $11.99**

The Cyanide Process of Gold Recovery - Unavailable since 1894 and released under the name "The Cyanide Process: Its Practical Application and Economical Results", this rare publication provides an in depth look at the early use of cyanide leaching for gold recovery from hardrock mine ores. This volume provides a reference into the early development and use of cyanide leaching to recover gold. **8.5″ X 11″, 162 ppgs. Retail Price: $14.99**

California Gold Milling Practices - Unavailable since 1895 and released under the name "California Gold Practices", this rare publication provides an in depth look at early methods of milling used to reduce gold ores in California during the late 19th century. This volume provides a reference into the early development and use of milling equipment during the earliest years of the California Gold Rush up to the age of the Industrial Revolution. Much of the information still applies today and will be of use to small scale miners engaging in hardrock mining. **8.5″ X 11″, 104 ppgs. Retail Price: $10.99**

Leaching Gold and Silver Ores With The Plattner and Kiss Processes - Mining historian Kerby Jackson introduces us to a classic mining publication on the evaluation and examination of mines and prospects by C.H. Aaron. First published in 1881, it has been unavailable for over a century and sheds important light on the leaching of gold and silver ores with the Plattner and Kiss processes. **8.5″ X 11″, 204 ppgs. Retail Price: $15.99**

The Metallurgy of Lead and the Desilverization of Base Bullion - First published in 1896, it has been unavailable for over a century and sheds important light on the the recovery of silver from lead based ores. Some of the topics include the properties of lead and some of its compounds, lead ores such as galenite, anglesite, cerussite and others, the distribution of lead ores throughout the United States and the sampling and assaying of lead ores. Also covered is the metallurgical treatment of lead ores, as well as the desilverization of lead by the Pattinson Process and the Parkes Process. Hofman's text has long been considered one of the most important early works on the recovery of silver from lead based ores. **8.5" X 11", 452 ppgs. Retail Price: $29.99**

Ore Sampling For Small Scale Miners - First published in 1916, it has been unavailable for over a century and sheds important light on historic methods of ore sampling in hardrock mines. Topics include how to take correct ore samples and the conditions that affect sampling, such as their subdivision and uniformity. Particular detail is given to methods of hand sampling ore bodies by grab sample, pipe sample and coning, as well as sampling by mechanical methods. Also given are insights into the screening, drying and grinding processes to achieve the most consistent sample results and much more. **8.5" X 11", 124 ppgs. Retail Price: $12.99**

The Extraction of Silver, Copper and Tin from Ores - First published in 1896, it has been unavailable for over a century and sheds important light on how historic miners recovered silver, copper and tin from their mining operations. The book is split into three sections, including a discussion on the Lixiviation of Silver Ores, the mining and treatment of copper ores as practiced at Tharsis, Spain and the smelting of tin as it was practiced by metallurgists at Pulo Brani, Singapore. Also included is an overview and analysis of these historic metal recovery methods that will be of benefit to those interested in the extraction of silver, copper and tin from small mines. **8.5" X 11", 118 ppgs. Retail Price: $14.99**

The Roasting of Gold and Silver Ores - First published in 1880, it has been unavailable for over a century and sheds important light on how historic miners recovered gold and silver rom their mining operations. Topics include details on the most important silver and free milling gold ores, methods of desulphurization of ores, methods of deoxidation, the chlorination of ores, methods and details on roasting gold and silver ores, notes on furnaces and more. Also included are details on numerous methods of gold and silver recovery, including the Ottokar Hofman's Process, the Patera Process, Kiss Process, Augustin Process, Ziervogel Process and others. **8.5" X 11", 178 ppgs. Retail Price: $19.99**

The Examination of Mines and Prospects - First published in 1912, it has been unavailable for over a century and sheds important light on how to examine and evaluate hardrock mines, prospects and lode mining claims. Sections include Mining Examinations, Structural Geology, Structural Features of Ore Deposits, Primary Ores and their Distribution, Types of Primary Ore Deposits, Primary Ore Shoots, The Primary Alteration of Wall Rocks, Alterations by Surface Agencies, Residual Ores and their Distribution, Secondary Ores and Ore Shoots and Vein Outcrops. This hard to find information is a must for those who are interested in owning a mine or who already own a lode mining claim and wish to succeed at quartz mining. **8.5" X 11", 250 ppgs. Retail Price: $19.99**

Garnets: Their Mining, Milling and Utilization - First published in 1925, it has been unavailable since those days and sheds important light on the mining, milling and utilization of garnets. Included are details on the characteristics of garnets, where they are found and how they were mined. 78 ppgs, 10.99

Gemstones and Precious Stones of North America - Leading mining historian Kerby Jackson introduces us to a classic mining publication on the gems and precious stones of the United States, Canada and mexico. First published in 1890, it has been unavailable since those days and sheds important light on the gems and precious stones that may be found in North America. Included are chapters on diamonds, corundum, sapphire, ruby, topaz, emerald, disapore, spinel, turquoise, tourmaline, garnets, beyrl, peridot, zircon, quartz crystals, feldspars, pearls and many others. Included are details on where these gems and precious stones may be found throughout North America, as well as their characteristics. 360 ppgs, 24.99

Mining Camps and Mining Districts - First released in 1885 by Charles Howard Shinn under the title "Mining Camps: A Study in American Frontier Government", this publication offers a unique look at how early gold miners established their own forms of representative government during the California Gold Rush. Drawing on the the early mining codes of mideviel German miners in the Harz Mountains, on the mining customs of the Cornish tin miners and early Spanish mining laws introduced into California, the miners established the first governments in the American West. 340 ppgs, 24.99

BLM Field Handbook for Mineral Examiners - Leading mining historian Kerby Jackson introduces us to a classic mining publication on mine evaluation. First published in 1962, this work sheds important light on the techniques of BLM Mineral Examiners to perform validity on mining claims. 132 ppgs, 10.99

Six Months In The Gold Mines During The California Gold Rush - Unavailable since 1850, this important work is a first hand account of one "49'ers" personal experience during the great California Gold Rush, shedding important light on one of the most exciting periods in the history of not only California, but also the world. Compiled from journals written between 1847 and 1849 by E. Gould Buffum, a native of New York, "Six Months In The Gold Mines During The California Gold Rush" offers a rare look into the day to day lives of the people who came to California to work in her gold mines when the state was still a great frontier. **8.5" X 11", 290 ppgs. Retail Price: $19.99**

The Discovery of Gold in Australia - First published in 1852, it has been unavailable since those days and sheds important light on Australia's gold mining history. Included are rare communications between British agents and the British Crown when gold was first discovered in Australia in 1851. This rare text contains hard to find details on Australia's first mining camps and Britain's early attempts to provide for the orderly regulation of gold mines in that part of the world. Also of interest are hard to find extracts of articles that appeared in the early colonial newspapers that did their best to report on Australia's gold rush as it took place.
102 ppgs, 10.99